SCIENCE AND MATHEMATICS OF THE UNIVERSE

Creation Law

KEON CAMPBELL

ISBN: 979-8-9933020-5-8 (sc)

Print information available on the last page.

This book is printed on acid-free paper.

Because of the dynamic nature of the Internet, any web addresses or links contained in this book may have changed since publication and may no longer be valid. The views expressed in this work are solely those of the author and do not necessarily reflect the views of the publisher, and the publisher hereby disclaims any responsibility for them.

Science & Mathematics of the Universe

Genesis

Creation Law

Black people are the original creation God commanded us to exist we now be it is what it is I am that I am made to give light in the dark of night God saw to it his creation was the only good us and we made I and me the devil being made us physical made us dumb and we unable to see are self which is most high.

I have karma I am here to burn in this life the reward of pass bliss as a king I am now a lord a God in the person as Buddha was a king made are bliss in the person he was resurrection as master ford came from the grave of lost and found flesh and bones ford was flesh of are flesh bone of are bone to have an ear to listen is to hear God is in front of the eternal God is so far back in time from nothing we can truly know and be comfortable without him there is nothing which comes from nowhere we are in the wilderness to feed are people

We exist in the Allah the body of Islam the good news is the creation law the only good the devil controls the waters the sky and the land the devil sucks the milk from the mothers he is a blood sucker of the poor as we know life by seasons dates and times how we know self and others will end it is time for the earth mother of us all to show and prove the creation law came by commandment of a said word let there be light and the Elohim became the light of a black God a God of divine knowledge which became a said word of wisdom the said word is before life as we knowledge it life of which the soul spirit that is too old to die one cannot die who was never dead light and life a mind of divine knowledge and supreme wisdom the sun gives light and life where there is light there is life yet we are far removed in our minds souls and the holy souls we know not self and others the heavens and earth have none to be compared yet the original man is the crown of creation a king known by his wisdom spirit is the nature of the physical matter spiritual wealth comes with physical wealth here and now not in the dead we look to the dead as adamites ate with the snake life is now not past or future now will always have existence were we see stars there is light and were we see light there is life Abraham was taught to count on the stars for guidance he was taught to return the dust back to where it came and reincarnation of the soul will become a soul of a holy

Stars of the universe and statues of dust were created to serve and not to be served look not to the stars in worship look to the dust to cultivate not as glitter and gold

The sun is healing, there is no light of the moon she reflects the sun and gives seasons years and days of which we cultivate the earth
God said
God commanded
God let
God saw
God said let it be good
God called out the creation
God gave rest to the scientist
And to the creation law they
gave there crowns

Daniel 2:32-35

Mystery schools' gold
Persia silver
Greece copper
Jew iron
Christianity iron clay
Islam clay
Buddhism stone

The Kaba stone will make us one kingdom in to one from the rock mountain of the prophet Muhammad and the angel Gabriel on the night of power the last days of the sabbath here and now Islam is a culture not a religion Islam will be made a new kingdom pilgrimage is a birth right of all nations the nature of a wise men is Islam made a religion by men of traditions that lead the pious away from there self who cannot see spirit is the nature of the physical were made dumb of spooks what we see is real what we cannot see with the naked eye we have science to see the nature of what we cannot detect the names and nature of the sun moon and stars are the nature of its cosmic space and the characteristics of its function by the age size and way of life the sun moon and stars are named the name above every name is keep secret the Allah is the proper name by the comparison with eloh in the Hebrew yet God is not the original name of the creator the Allah is the body of Islam is a name we only know the wellness and well-being of the Islamic body will transform the mind brother Jesus did not leave the message in the Quran he was by God a example of the Mahdi who came to the lost found in the person of God the law of creation is change cosmic slop atoms cells molecules the preserves of the universe came before the sun moon and stars God is the light the awareness of the universe that cannot die were there is light there is life

monotheism is in its origin is the aton the sun disk of Egypt the rays of the sun disk hold the ankh the key to life as a people of one God though out the original black history we were a people of exchange knowledge and resource though out Europe and Asia what is known now was known then the Elohim are the fathers of Abraham Isaac Yacob the God of the masonic law is dead with the Jesus are master ford the person of God the twin spirit of are brother Jesus the masonic God is Yahwah I am that I am he is who he will be Daniel 12:2 he is light and life to some darkness and death to others the scripture cannot be broken in the beginning groups of supreme beings created the heavens and earth these groups are Gods sons of God angels the masonic law is dead are rebirth is in the creation law from soul to holy soul God is a God of justice positive or negative good could be less evil could be worst all ways reflect on life what happens good or evil could all ways be worst we must reflect and purify here and now the saving grace is now prayer which is good yet God consciousness is the best form of worship

genesis 2:5-6

as by the cultivation of the rich black soil of earth gave way to rain one of the reasons we are in tune with water

works of the woman he in her womb of waters we culture the woman as we culture the earth her twin spirit we have a duty to divine knowledge supreme wisdom and overstanding we have a duty to learn the sun moon and stars we have a duty to man woman and child there was no jew in the genesis creation law there is no jew in the torah masonic law the jew is the crucifix of Greco-roman horizontal judeo-christianity vertical the good news of Islam through Abraham a message of north east west and south a circle of wise men of who came the prophet with the gospel of the said word until the person of God the iman of the world who has come the al Mahdi from the holy city of mecca the birth record of creation has no beginning or end the study of the creation law is fulfilled by brother Jesus the creation law is a stumbling block to the masonic law of the jew the fact is that the creation law could not be fulfilled until brother Jesus a sign of the last to come the one of which the prophet Muhammad gave word of the last to come the al Mahdi to a lost found Abraham was humbled with the surname ham he became the father of world religion out of Egypt we have come are food stuff genesis 9:4 eat not flesh it is a living being of life blood milk

water

wine

honey

fruit

vegetables

fish

raisins

brown rice

beans

nuts

dates

bread

The 5% of the earth teach the poor in mind righteousness who be the 85% the 10% are rich of physical wealth the 10% by wealth makes the 85% mass uncivilized poison animal eaters the 5% teach with mental wealth there comes a physical wealth all is finished in the Jesus by the creation law he is here in are day and time master ford malachi 4:1-6 the Quran is the cure all a furnace to some a healing to the God tribe of shabbaz the sun in the west a law like moses given by Elijah from the person of God being the divine knowledge of self and others the supreme wisdom we have been made the son of man the sons of God when we are the mothers and fathers of civilization God in the person they will say they are our people we will say jew we are Hebrews just like you we will say God is mine not yours they did not know God Deuteronomy 30:12-14 there is no holy outside of the said word of God and a firm heart energy cannot die self preservation is the first law the heavens and earth will change shape and form before a end the brain is a power plant of energy the more information the more energy the more one observes the more one reflects the more one takes notice the more ignorance the less energy and you die in the thought which is the most valuable resource of the brain the most valuable organ of the body is the brain the muscle we exercise by thought you must observe before the said word can flow from your lip like water divine knowledge before supreme wisdom foods and necessary waters say not 3 yet God does what he wills if you know the system as the devil you must know the monetary laws which is where we find social equality a equal footing to be ignorant is to be of a mental poor the devil is of a filth his wealth is the root to all evil we as the original men and woman are a people of resources we have always been known by are fruits the parables of the Jesus are the said words of the creation law of genesis the said word of the creation law is in the heart not hard or soft it is firm Jeremiah 31:33 the heart is law some are broken in law yet God cannot be broken those of the commandment let there be light will always be Elohim Gods who gave the light to the kind of man Daniel 12:4 the earth is in a new sky house of information these days said word is fire the rainbow is not a blessing it is more a warning Islam is a culture a way of life Islam is science and mathematics of the universe Islam is a natural stimulate a culture of sincere men and woman

Observe
Reflect
Purify
Ponder
Measure
Weigh
Harmonize
Salat prayer
Dua prayer
Dhikr prayer
The lords prayer
The buddah pose
Lutus pose
Breath control
Mantra
Karma
Karma sutra
Cause and effect
Reap and sow
Synchronism
Martial arts
Exercise
Art
Music
Hail marys
dance

God is honey to the lips a sweet kiss butter on bread a tall glass of water a glass of wine pure milk of the cow as the hindu and the Egyptian the masonic law cannot atone yet the creation law is here and now prophets were doctors of the masonic law the coming to atone what was lost and found by the 144,000 a small percentage the God tribe of Shabazz the 5% nation of the Gods and earths look listen learn observe with respect to God self and others the physical eyes see what appares on face value the reflection of the 3rd eye makes things clear God was made man in the mass man became God in a small percentage the 144,000 5% genesis 3:16 the womens emotions are labor pains body and mind planets and stars sit as seeds in fresh firm darkmess the lamp of the body is the mind the earth the women travels by her man the sun the said word gave way to the creation law so called angels in the likeness of God were made man under satan the Jesus is the flesh and bone out of Egypt are bread and wine there is no religion that can be compared to the culture of Islam the world order is a one world religion created and made by the 10% rich who control the mass by the control of the waters the sky and the useful land we are made a mental and a physically poor made slaves to there wealths Abraham came to the canaanites moses came to the Asiatic egyptions Solomon came to Judah by the captivity of the kings came the jew a salt of Hebrews and isrealites what good is the jew if he has lost his flavor what good are we by the creation if we as a people have lost ares by law we are mind and holy soul to the Allah the body of Islam the creation law was doctored by prophets to no end only to fail the arab is Nubian by the matriarch hagar and the father Abraham the baptism of God is the conception of seed the supreme mathematics and the supreme alphabet and the 120 degree supreme wisdom the universal flag and earth lessons the measurements of the heavens and earth the measurements the weight attaction and speed she is the earth lesson not found in its proper text in the bible or Quran her lessons complement the bible and Quran the basic instruction before leaving earth the bible the cure all the Quran ezekiel 37:17-14 the mass is of a dumb a mental dead to a mental grave of dry bones yet by no set time the devil must let go of are bodys and minds my knowledge is her wisdom my wisdom is her knowledge her weakness is my strength my strength is her weakness the black woman is the earth that holds the roots of all nations all people she is the mother of civilization knowledge verse wisdom wisdom verse knowledge man and his woman must compete to righteosness you are the universe the universe is you the eggs of the womans womb are planets she holds a universe of egg man holds a universe of seed seed and egg storms in the womb in a constant fight for supreme

7 senses

Eye light
Ear sound
Nose smell
Mouth taste
Brain mind
Heart emotion
Sex organs baptism

Man is the root woman is the earth that holds the roots in place u-n-I you are the universe the universe is you genesis 26:5 the laws of Abraham were given by melchizedec before the masonic law was the laws of melchizedec a coming from enoch who walked with God above and below genesis 6 : 1-2 how can God not have offspring when he has so call sons heaven and earth are subject to the creation law God is a man with knowledge of him self know the ledge and fall not ignoring the consciousness found in what you know supreme wisdom is a wise dome once you wake up to what is yours you can wise the dumb here and now the mental is the essence of the physical as fire burn come smoke the devil knows not the mind in his study of are old works as original people the devil study monkey and man now he study man and God the sun moon and star and the solar boats being above and below the mother plane of egypt the nation of Islam has no birth record yet it was made a religion which has a beginning and end Islam is life life is mathematics and the light of Islam is science and mathematics of original people Islam is culture the black man is law the true way of life is Allah the body of Islam

A king is by his kingdom wise only Islam is a life worth living you are a savior of self and others self is one who has the powers to take self from the depths to the heights from heights to the depths the mass knows not the trust is the law of the black man the question mostly asked by the mass is why there is no question the God cannot answer power is the truth the magnetic ability to attract the positive energy in you being God a queen is one who travel by the speed of her God u-now- know
The unknown the more the devil experiments on the sun moon and star he is sent back to earth or his like of man and monkey no nation will see a divine destruction before the karma here and now to grab and rise the lost and found the universe is you the black man is law we have won the victory by the wisdom of the wise we are physical yet the mass needs the mental we have found the body yet we are lost in mind wisdom is the wise words spoken by the wise black man to show and prove the black man is God by law the Jesus came to fulfill the creation law from a dead masonic law of kings and priest of Israel the black church is a village of elders a tribe of fellowship the minister the chief yet they know not

We are living in a day and time of reminders all throughout the globe of a hell we are living in fire now to reflect on what are the happenings of today is the best prayer in the masonic law the dumb animals are given for the atonement of man the Jesus cannot save you you are the only savior by self now is the time to built basic instructions before leaving earth the b.i.b.l.e. the holy cure all the holy Quran the Jesus gave the law of creation to the man of the original law the parables which cannot be broken being Gods by nature and person the lost found life by the creation law not the dead masonic law do we have a creation Ezekiel 1:10

Man universe
Lion sun
Bull earth
Eagle star

Exodus 1 :15-29,12:29,13:1-2,21-2 the first born was the Hebrew the son made slave by the Asiatic Egyptian government made blessing the mother and son over the father this blessing of mother and son predicted by the doctors of the masonic law from where we fell from gold of the prophets the doctors of the masonic lost to the kings and priest who see by face value physical wealth the root to all evil sown in earth we must see clearly what is and not what appears to be the black law by the creation law the germ of the black man was taking yet it is still are speed and power over the devil to be made weak by every step in the creation law of red yellow white germs the black man is law the true way of life is Allah the body of Islam God is known by the wisdom of his kingdom his knowledge of self

We must work in order to eat we must divide the fruit from the weeds the truth of Allah the body of Islam

to slave ships we are the breath of God the holy spirit in the bodys of man Allah the body of Islam the magnetic ability to attract positive energy is the power of truth which makes a people free God is not invisible he is hardly detected with out reflection by which he is seen and heard
aries ram
taurus bull
gemini twins
cancer the moon
leo sun
virgo virgin
libra weight scales measure
scorpio sand
Sagittarius the horse
Capricorn the goat
Aquarius information age
Pisces the two fish

Wealth is the root of all evil we are given glitter as if it was gold we must not be evil with wealth this is the devils work we can have spirit wealth now in life or in death and physical wealth in this life and be burned with karma in are next or we can do with out in this life not by choice we are poor as a thought we all came to the life as we now know Daniel 12:4 this is the information age the signs are bright as day Islam is the best advice given the uncilized moses was the child of thought thuthmoses I II III raamses child of ra [the sun] a muslim is one who has given his mind freedom from religion by choice there is no force in Allah the body of Islam it is a culture of community sense and sound judgement we are free to prove God as we know him in are person Africa has the resourecs to rule the earth slaves were income to satan we are still cattle in mass 10% is law given by moses the Israelite lost all knowledge in Egypt the house of bondage of pharaoh and priests of a true and living God eternal thought out creation not a mystery God unseen yet everywhere is God Israel went savage and was given the beast to atone for there souls under God yet the Jesus gave us a pass of fulfillment for these last days the devil came with light in his beginning now is come his dark day

Summer sun
Fall earth
Winter universe
Spring stars

Creation has a birth yet we have been here so long we gave witness to its coming the original man cannot be found yet he in bible and Quran was the word be the word and then God and God was with the word and the word was God wisdom wise in the mind or dumb in the mind freedom free in the mind or free to be dumb are those who know not the word is God heaven and earth are subject to the original man who is the creator of the creations creators God is man with knowledge of himself who knows the ledge and stands firm on the facts as far as he is learnt yet the fact that is mine is I am God because I look to my self the foundation of all my information is infinite the fact of all exist in the known to be found and given the mass the fact knowledge is infinite is the fact of God himself

The life we are living which is the Allah body of Islam is the culture of science and mathematics the culture of Islam is life and life is living what you cee as the best part of your life live it out your freedoms and powers are rooted to your earth in equality yet she cannot pass her seven her God is supreme in his being given her her speed she is 24 day her God is 24,000 years she is under the sun she travels by the movement of her God at todays set date of being the last the sun is at its most fire we all cee yet the mass being blind deaf and dumb have keep the devils happy to long Yet today we are all jehovahs witness who must reflect and refine the word of God in the mass the 100% being all peoples of today the 100% of the 5

The sevens of revelations
The seven lamps
The seven stars
The seven churches
The seven seals
The seven trumpets
The seven beast
The seven images of the beast
The seven plagues
The seven bowls
The seven angels

before existence we were here
before the heavens we were here
before the earth we were here
before the works of old we were here

the 7 chakras of the body

1 the crown
2 the third eye
3 reality
4 truth
5 solar plexus
6 Astro plane
7 Physical plan

99 attributes of Allah

1 God is all mercy
2 Gods mercy is most best
3 Gods dominion is his kingdom and ownership
4 God is most absolute
5 God is the giver of peace and perfection
6 God is the only one who gives faith and security
7 God is the guardian overseer of all
8 God is the most in might
9 God is the restorer by force
10 God is the supreme majesty
11 God is the creator by making a command to be
12 God is the origination
13 God is the fashioner
14 God often forgivings all
15 God when he subdue he is forever dominant
16 God is from where all gifts come
17 God is provision
18 God will open all to judgement
19 God knows all and everything
20 God withholds yet he has much to give
21 God will extend
22 God will reduce and humble
23 God will exalt to elevation
24 God is honor bestowed
25 God dishonor those who humiliate others
26 God hears you
27 God can see you
28 God is a judge of the best judgement

29 God is just to all

30 God is gentle with the vain

31 God is acquainted with the aware

32 God is the most comfortable

33 God is a supreme kingdom and lordship

34 God is a abundant in his forgiving

35 God is appreciative for and by worship

36 God was made most high by those he has exalted

37 God is the greatest grand architect

38 God is the preserver of all who have his protecting

39 God is the only

40 God by time is reckoner for those who know not

41 God is majestic

42 God gives out of respect

44 God is the most responsive

45 God is all encompassing of what is boundless

46 God is all wise

47 God is the best lover

48 God is a glory at its most honorable

49 God is the resurrector

50 God is forever witnessing all

51 God is absolute a perfect and pure truth

52 God is trustee he gives by conditions

54 God is firm and unchanging

55 God is a protection a associate to lean on when in need

56 God is worthy of all forms of worship and praise

57 God has a numbering and counting of all in a record

58 God is the originator the initiator of the creation

59 God is the restorer and life ls made reinstate

60 God is the giver of life

61 God will destroy the bringer of death

62 God where ever he is he is living

63 God upholds and sustains as he will

64 God is the perceiver who has God has the holy soul

65 God is a well known magnificent king

66 God is one as in first one as in ranks

67 God is unique in his creation of himself self and others

68 God is the eternal satisfier of all needs

69 God is most capable in giving power

70 God cannot be won over he is the almighty

71 God makes what happens to promote him self

72 God delays sickness

73 God is the first the sole one in the ranks

74 God is the last to create him self

75 God gives manifold

76 God knows best what is hidden

77 God is a governor who gives and takes not

78 God by self is exalted

79 God is the source of all good his giving is in happiness

80 God frees us from harm

81 God gives punishment in return for injury

82 God has a pardon for us in grace

83 God is a friend most kind

84 God is the master of kings owner of the universe

85 God his glory and honor is majesty with a open hand

86 God is fair in all his dealings

87 God is the gathering of the unified

88 God is wealth he is the resource to go for help

89 God is the enriching of the earth for are betterment

90 God gives not in full he withholds

91 God is in control of harm as he wishes

92 God is in favor of and for the right cause

93 God is the light the illuminator of life

94 God has made the stars as guide to him self

95 God is the only orientation to freedom

96 God is surviving in a everlasting creation

97 God is the inheritor and heir of the dead

98 God made stars which guide God is a error free

99 God is most patient

We are created by a self created God who created us out of him self we are him he is us we are with him he is with us to be is the commandment of from where came light and were there is light there is life the son of man gave eye to the coming of God in the person before there was a creature the son of man was here self created in God as God was self created seen and heard by the son of man once you wake up to what is yours you can wise the dumb to a state of supreme wisdom which is the science and mathematics of are culture which is Allah the body of Islam light is science and were there is light there is life being mathematics the reward for embracing are culture is love peace and happiness the penalty of living in discord is hell the original man is spirit placed in dust adamites was taken from the dust of there true paint they were the first believers the first prophets the root of the original man is spirit the tree of life the root of adam is dust the tree of knowledge of good and evil is from the devil came from the original mans flesh is the flesh of adam the original man is spirit of spirit the flesh is weak yet the spirit is willing we are living in a day and time when prayer is most needed God consciousness is the best form of worship in the masonic law the dumb animal was given for atonement yet that is zero that has no add of knowledge

Prophets the doctors of the masonic law tried to atone only to fail the kings and priest made the masonic law fit there devilish deeds man is the root woman is s earth that holds the roots in place you are the universe the universe is you

U-n-i by the creation of the moon by a God scientist the pyramids by the pharaoh colors of man by the scientist

The God Yacob these are wonders of the world yet the son of man was created when God created himself before the sun moon and stars the wonders of God and his science and mathematics the light and life of the son of man God is a man with knowledge of himself know the ledge and fall not ignoring the facts the culture is older then the trillins of years of stars which is in are space the son of man planted the fruit of the tree of life satan planted weeds the tree of knowledge of good and evil the plan of God was to let the fruits grow with the weeds to a set date so no fruits will be lost the knowledge of good the weeds were the knowledge of evil adam was made to work the weeds from fruit in sweat for eating with satan only Islam is a culture worth living now is the time to be positive or we will end as a nation to have insight is to comprehend by the minds eye to see though the physical eye to understand the 3rd eye no nation will see divine destruction before the resurrector of the lost and found devils are leaders in evil they plan day and night yet God is the best of planners you are your self a savior one who has the powers to take self from the depths to the heights back to the depths a queen is one who travels by the speed of her God rule with her self is her only savior the future of the nation is in her womb wisdom is the wise words spoken by the wise black woman to show and prove she is the earth lesson it is not all about sex yet the karma sutra positions were not made in vain we must respect are woman maintain there happiness examine them support there faiths lean on there understanding the original man is spirit which became physical matter we became trapped in physical hellish flesh the spirit is heaven flesh is hellish weak spirits must be tested the spirit of God is not to be put to the test the spirit of God in man is not to be put to the test the spirit of God in man is willing the nature of the flesh is weak the spirit of God in us is not to be tested what God wills is the spirit not flesh by the fall of the spirit in to the physical came hellish worlds the moon is the queen of heaven the kings mother must be honored a counselor to the king are soler star the messiah seen in the sky of his coming among us were there is light there is life the physical body needs head and mind the creation is the reflection of information the original man is the authority of heaven on earth the moon reflects the information of the sun the earth grows knowledge to a understanding child and star the original man was in the world as one and all else is secondary to his existence it was by with for which all things came together and apart from him the spirit of God would have not have gave command to let there be light one who knows and declares what he knows is one of wisdom one possessing much knowledge God did not end or stop creation we will give the devil no rest from the wars of heaven high desires are God him self low desires are devil the earth is a constant growth and development we live for ever by the mind yet by karma we all must taste death this is the reasoning why we live to learn and observe the past were from you come we were never dead we were here before the womb we will be here after death to see God with in side is inspiration moses seen not God in him self as God commanded a cursed leader a cursed people those who mock God will come fire and floods moses in a circle of fire in a mist of cloud 40 days as the christ in the wilderness God said let there be light yet some of the heavens broke out in war the state of devilishment devils are in high places hills and mountains there is a war in the clouds the high give comfort the war is not flesh and bone it is on high above messengers as enoch yacub [Jacob] moses the Jesus prophet Muhammad all made way above to fight hellish worlds in the mind spirit and soul all being the knowledge of self wisdom is the battleground understanding is to see your self win the war Islam is not a religion it is a culture a way of life by which mathematics and science of the universe we apply every day to are circle God in the arm leg leg arm head Allah the body of Islam is not found in the church the synagogue or the mosques the physical flesh bone is the temple of God Islam is the nature and essence of God him self the baptism of Allah is the womans labor the original womam is mother of all living knowledge is before wisdom many are called yet few are chosen some get information and have no wisdom which is a reward yet some have it not this is why we are Allahs 5 percent nation Abraham was the good news Abraham saving the two Sons the Hebrew and the meccan there is no new the saving of the only son out of respect were came the Jesus his people call them self christ the words of prophets are dead in this day a divine mind soul and spirit is God him self doubt not test not the God in you say not trinity yet God can do as he will woman is a limited example of God yet she is queen over prophets in this day being the last as we knowledge God is not a color black is a state of mind black is so old in time in space there is no evidence of light before heaven and earth were split black is old light is new to existence black has no beginnings sun moon and stars are new to exist black is so old it is nothingness the original man was here before light before stars his culture was in the Old ways and actions with out count of time when the light of planets work as clocks the sun is a star they are also planets as all other space bodys are planets the sun shine on the good and evil it rains on the good and evil we are all his children yet a few are chosen thou shall not lust gaze stare or give the evil eye to those who are called yet not chosen we are all children of heaven the jew is a chosen CaucAsian they are both gentles yet the jew is the chosen of are people the CaucAsian a race out of the cave devils have no part in God he fathers the righteous by master ford the message of Quran and bible in his person to the lost and found tribe of Shabazz who call them self christ

God
King
Lord
Master

I give my life for my culture

i God

The woman travels by the speed of her God we are slave to time we are measured giving space and movements some cannot see from were they came before the light of the sun before mother and father they forget God gave us these days s become no not karma it is not what goes around comes around karma is life you cannot kill what was never dead the mind cannot die of its information each incarnation is a opportunity to learn and observe your past your now and the last day when we see the bliss of God karma is constant are day is thousands of years to God are day is 24 hours years by the circumference of the earth with God are day is 24,000 years by the measurement of the universe time is the master of illusion some run from there karma yet the fire of karma is given as to learn and humble you before God before dead we are the dust of earth which we must return and live out are spirit with Gods the original man is the father of all information his wisdom being his woman is to show and prove her man is her God do not put God to the test with questions look to him for answers call it be and it is God willing to be is the commandment God gave his angels as we call on him the answers exist they are throughout the universe

Observe your a-alike
Reflect who they b-alike
See a-like who see a-like you

Awareness is meditation a form of prayer God is the rule with out him there is none magi the magic of yin yang black and white magic is the mathematics of magnets that attract by there positive energy and comes a positive and negative charge in order to live out are good karma we must be wise to negativity and build by this we will live in a existence of no end by Gods time God knows are condition hears are cry and knows are need God is the one supreme God is the keeper of the best God is the most merciful of a-alikes to those who show mercy God is most a-alike God is the black state of mind from were came light which was broken in to colors which became angels who came with time yet the Gods have no birth record they are so old God is the father of colors we must raise the fallen angels up from there filth they are not spooks they are physical men and woman we are the black state of mind from were came colors which is broken light woman has no light of her own the sun provides her labor by a secondary light the moon that reflects and give growth to earth which is infinite in his eye the supreme being has knowledge of no equal we must work are karma as cultivation of the rich black soil from were we come from as black people being black in state of mind we must make effort to sow positive karma we must move negative karma with good works Gods pleasure is to make you happy with the best of health man is a false face we are God in the person man is lost yet we must except those who are most patient and persevering being the woman in labor God was present with master ford who gave the supreme wisdom to the lost and found tribe of Shabazz nation of Islam the Allah body of Islam

21

Allah 5% nation we must understand more then we know wisdom gets us to understandings there is no purpose other then God he is the master build we are the master builders we must work are

the brain muscle
the third eye
the vocal cords
the heart muscle
the gut
the navel
the sex organs

in each person God is present good or evil
in each family God is present good or evil
in each race God is present good or evil
in the earth God is present good or evil
in the science and mathematics of the universe God is present good or evil

God is the best at plans when he is the goal all is possible God is said to be unseen in the mind of man yet God is seen and heard everywhere nations die for the transgressions of there leaders rule they no not the light of knowledge the rich black firm soil of wisdom the seed of understanding and the peace that comes with the cultivation of the mind the brain muscle from the darkness of soil comes the growth of the seed to the light of life from darkness comes light from light comes darkness

God is
Vast
Immeasurable
Broad
Wide
Large in scale
he is heaven it self

God is mighty by wisdom creation after creation in karma everything will pass yet the judgement of God is above all judges to him we are constantly brought back we grow and develop there is no knowledge in the grave we must get wise to God here and now information is elevation it is like the air we breathe God knows the milk of the breast pure water to wash and clean the queen bee of honey the vine of wine christ the son the earth is being observed from above are heads the solar boat of Egypt the wheel of prophet Ezekiel the powers in are inner are atomic we are nuclear when the atom is split it holds the force to split heaven and earth many times over these powers have forever been with the original family man is a false face the original man is not man mankind or human before the false face was the original God Allah the body of Islam arm leg leg arm head supreme man is of his last day we are near to him then his life vein the vision of the prophets comprehends by the doors of the two eyes yet the original man is of all vision by the 3rd eye to hate is to make enemies hate not those who hate you kill them with kindness love is a old testament law the old testament is dead yet 0f the old testament we need the fulfillment of the 7 day creation law which is the geno the genesis and the exodus where we exit the dust not the mosaic law the Jesus came to teach the jew of the God of all man and woman God is not alone theres a God of all nations not a God given worship for wealth and resource out of greed knowledge is before wisdom one must know and reflect it to wisdom to be a wise God one must show and prove the black man is God to his earth knowledge is before wisdom God comes with the growth of knowledge in wisdom the rich black soil understanding rotates around the sun in every direction man can rise on high by his own effort the magi were a host of angels from what is now iran the lost sheep of israel we all have a personal God some cannot show and prove who they think God to be which is the duty of the 5% nation those as the magi who study the star of the Jesus are rightly guided as Abraham by the study of the star he was given sight of his seed yet many were called and few were chosen those who see not the time for universal cultivation are in the illusion of time and hold no place with the will of God we must murder the devil in which the mind creates or we will be punished he cannot be reformed so he must be ran of the earth with no lost of time God in reverse is a dog to live in reverse is a world of evil wisdom cannot come before knowledge which is dog the reverse of dog is a life to be lived for and with God

I self
You
We
Us

Keon Campbell

Me

u-n-i

i-n-i

the universal rotates around the Asiatic black man in 8 direction to hear clearly is to see his face not the false face of man one must not and cannot turn from us she sees us with the reward of unconditional love life or death the jews make apes and swine of the people in there teachings the Jesus made the deaf dumb and bilnd wake from there sleep knowledge will come to the dark places of wisdom and some will wake to light some will wake to darkness some are hidden from them self and cannot see yet there are those who see what is hardly detectable and see the science and mathematics of Islam we are due and deserve God in the person of are being this is the only freedom which is are justice for are labors under the slave master and his children a complete freedom is in you created in side you focus on the in side and create a world on the outer limits of the mind with out fall every action is after a thought which in the physical existence up root negative energy and plant seeds in are circle by positive thoughts put in action light is the first law God is the law before all the command was to be as is so will come to be as is by God creation as the heaven and earth were broken apart as women is broken in labors what we think life is in karma there are returns back and forth zig zag zig the resurrector is God the return by knowledge is infinite information it is old before creation to God is in front of angels God is in front of the Gods God is in front of devils God is in front of man the woman is the birth mother of the universe helpmate of God given adam CaucAsians wish to live there 1000 year sabbath out of the abyss to snake the original people who are Hebrew as they are jew yet there time is set heaven is are mercy earth is are compassion as the lost and found earth follows the heavens from mercy comes compassion earth follows the heavens night and day as we enjoy God we forbid evils look not to your own self gain yet give to gain of others this is the mind of the Jesus who seen not vanity in being equal to God in all man God brought us from the wombs of are mothers being the helpmates of God she was given adam living creatures are through out the heavens and earth some we know some we know not the Jesus came to us and we received him we are his people he came to us and we hated him and his works being that of deity a human yet works of God the wine grape from the vine of the son of Solomon Menelik 1 the lion of Judah we are forgiven by the Jesus work yet we must find comfort in are mind and body by the life we live being a constant war with the beast word is bond and bond is life we have mouths to feed or we will be punished we are the children of angels the mother is embedded with the seed of angels who which we are superior to the angels the Jesus is above moses all things are constructed near God rewards come with righteousness punishment is the curse worship prayer faith belief hope all have conditions love fulfills all these being unconditional the lost and found in the ignorance with out a messenger as the muslim until master ford gave us purification we are told in holy Quran and bible we are a people to be given reward over all earth a small percentage from darkness came light to the muslim and that light became dark in todays day the dark days will come not understood by man yet the original man will know and reflect the cycle of infinite births and deaths will end and we will meet the bliss of karma in the state of resurrection we need God of most so there is a duty to have mercy and grace God is a mighty wisdom we are Gods by the commandant to let there be creators of light in order to put the heavens to work waters above were split and came waters below and came earth under the sun moon and stars from were came angels man came from the earth angels were sent down to earth from the heavens and man became Gods in the likeness of angels by the seed of angels and the daughters of man light was made in order to make what was hidden in the darkness appear the original man is the crown of creation the brain muscle and the 3rd eye clot some of those around you are of them selfs and there needs and wants or a savior of self and others self with out knowledge of self and others are incomplete in self is I yet the eye must have something to see and that is a civil self with the ability over others who have not the eye to see what is hardy detected by the naked eye yet is clear in the clot of the 3rd eye by righteous men wisdom was before creation and used by God to create the universe Islam is not a faith it is the science and mathematics of the universe devils cannot be of a teaching of Islam by faith or the sword the black church is a tribe a village of meek and humble followers of the Jesus yet they are not known to self and the teaching of the Jesus is not full in there hearts and minds the black church needs to break free from the beast which is the CaucAsian christ he must be circumcised from heart this is the sin of the flesh that must be cut are mind is white washed we are taught far from who we are sin is the flesh when the mind does not do its work we were made a stump yet the root cant be pulled from the earth we build life in order to destroy death we must react to negative karma with good deeds positive karma has a positive response on the negatives karma is born with us to live is a resurrection from were we come are pass life is why we are living why we will continue to live there are stars above which are dead yet they are still seen this is a example of the resurrector the original man of God those whose good deeds that are heavy in brain weight are the successful God is love and love for the word is love for God

We resurrect after we taste death karma is the fruit of life the original man is the resurrector God incarnate proof seen and heard can be proven by no set time we are here for answers not questions the hour and time is for listening to what is seen and heard is not to ignore or forget the earth is in a new space aquarius as the mayans predicted the so called global warming we must reflect and grow by what we are witnessing day to day to be aware is better then prayer children are taken to God young he knows the pain of there future he takes them to him self out of mercy and grace

Some of my inspirations

Noble drew ali
Master ford
The honorable Elijah
Anwar sadat
Muhammad ali
Yasser arafat
Malcolm x
13x father Allah
Khalid Muhammad
Marcus garvey
Dr york
Dr king
Haile Selassie
Minister Farrakhan
Gandhi
Nelson mandela
Huey p newton
The dalai lama
The buddha
Mansa musa
Genghis khan
King tut
Confucius
Moses
Menelik 1
Prophet Muhammad
The Jesus

To fulfill the creation law the Jesus was sent to a dead world he is the light to all populations yet many are called to God him self yet few are chosen of the black mass the gentile is CaucAsian a small population of jews are chosen gentiles these few CaucAsians teach and study lies about a unseen God yet the 3rd eye works by observation of the brain which is a muscle which is made supreme by thought mind and mental we observe God in the brain what we see is real and we must respect his command to subdue the earth devils came and made subjects of the black mass who no not them self they will not let the few chosen in the black mass the jewish gentile has the black mass dancing singing acting a fool for a doller dealing in play and amusement gold for glitter God which we are made blind deaf and dumb to his existence so they say he is invisible he cannot be seen yet that would make him a lesser of a God and we know God is most high in the ranks he is the potter who molded us in his likeness he is Allah arm leg leg arm head supreme all is real nothing can come from nothing Allah is the body of Islam the God body Allah has no birth record which is the illusion of time in the sun moon and stars a day to the average man is 24 days are time table is 24,000 years a day by the Asiatic calendar by which we know not as was first known by the genesis creation law a day was evening the a day was morning shabazz was to make a new man yet he could not make this man in Asia this man could not be made by these high sciences so this man of Shabazz was made by nature in the jungles of africa yacub [Jacob] was predicted a people of high sciences under law this is the color man yacub grafted devil

1 communication
2 technology
3 religion
4 theory of knowledge
5 perception
6 philosophy of science
7 metaphysics
8 language
9 power
10 mathematics
11 probability
12 arts
13 food
14 justice
15 bioethics
16 moral psychology
17 literature
18 choice
19 meta ethics
20 ethics
21 law
22 theory
23 biology
24 health

Shabazz in the jungle and his tribe gave names to animals by there natures in sun air waters heavens moon s stars chaos the heart balance fertility the animals natures the tribe of Shabazz studied were the strength and ability over predators a mothers protection discipline 24 sciences of Shabazz in the jungle were natures of animals the 24 sciences in the jungle was how the nature of animals are to be universally known as the powers of the pharaohs the creation and creatures

1 lion
2 hippo
3 cat
4 pig
5 frog
6 dog
7 bull
8 cow
9 calf
10 hawk
11 falcon
12 baboon
13 ibis
14 cobra
15 honey bee
16 scarab beetle
17 scorpion
18 crocodile
19 vulture
20 goose
21 phoenix
22 owl
23 leopard
24 quail

Shabazz was before yacub [Jacob] yet the man of yacub was to be made in order to be ran across sands to show the original mans power over the grafted snake Yacob the egyption held the snake as a sign of Yacob and his made devil Yacob was a God yet he was the father devil the color man is a hu black is not a color it is the state of mind of the Asiatic black man Yacob found the germ in the sex organs when the sperm hit the egg the egg took on a electric charge some call spirit Jacob found out the sperm had a unalike attract and the egg repel it had no attraction there are two germs in the original man a black germ and a brown germ the brown babies when born were made holy the lighter they became the black germ is most dominant the sperm with its attraction powers as long as Yacob could control the colors of the babies his idea would be born yet Shabazz was in the way the original man in so called African yet we know the east as Asia fire to wood is karma good deeds burn evil actual facts and truths are the reality we live the slave master gave us a mystery God we are deaf dumb and blind to the faith in a invisible unseen God we cannot see until are karma is made bliss or are resurrection in the flesh by the mind here and now is the first resurrection one cannot see the second resurrection if they pray to and worship a God they know not in this life how can we know a dead God that will save us in a heaven we cannot see some look up some look to a holy land they are taught to have faith in what is seen as evidence of the unseen we see the God body is arm leg leg arm head God Allah the body of Islam this is self knowledge seen and heard everywhere there is no heaven out side of you 7 universal centers of the body inside not outside the body this knowledge is solid this wisdom is life swift and changeable like water which is up from earth back down to earth in karma we come to earth as seed from past karma we past into a new karma as elders when we go up from the earth to return back as seed cream rises to the top a understanding is not just to see it is to be clear in information and the wisdom of your words due to knowledge we have wisdom self or savior only self can save yet to look out side of self for a savings grace to provide you with food clothing shelter we in the 5% say we are the small mass who must recognize in this savior who must knowledge self as the person of God if the messiah is not acknowledged by the 5% this is how we are to know self in the savior Islam is a culture the science and mathematics of the universe it is not a religion which is advice false knowledge is more dangerous then ignorance is a form of oppression which is worst then death we are ignorant by are oppressors made dead in the mind my ways and actions are mine alone my mind is mine God is guide in thought planned out the books of religion are advice science can be proven by facts truths and reality the sun moon and stars work by science night and day we see the work of God and angels we see more with the 3rd then the advice of religion if the trees were pens and the ocean was ink the creation could not fit in a book form this creation is made by the science of art reading and writing the mother of all books is open in what we see there is the mother book you are your universe your universe is you God revelations is the white hair red eye and feet like copper the helpmate of God was a woman sun and moon under her feet she was pregnant with the son of God the Jesus we all have a helpmate being the holy spirit God has come in the person in these last days we worship God alone yet we pray for many yet God holds his weight in worth in his observance God paints pictures we are works of art woman is a feminine the expression of her God

When you manifest knowledge right and exact night and day consistently in all your dealing the woman who observe her God in his kingdom is queen a true queen knowledge her God she travel by the speed of her God she is 6 and cannot pass on to 7 she follow 6 by her limitions 7 is her God he is the supreme being she is his helpmate to be married is heart to heart rings a verbal agreement which is legally binding abort your babies not use your mind to abort is murder from the days of moses there has been abortion which by the CaucAsian came out the caves and the land of Palestine and mecca Egypt and Ethiopia [utopia] moses was sent to Palestine to civilize those out the caves which before was a land of the calf and his milk and the honey bee the calf was yellow in color not dead gold it is like the red cow moses broke the law of the calf yet the law was rewritten for the red cow symbolic to woman she will not allow any filthy ways and actions around you and the child the heaven you have made together the brain is a muscle that must be cultivated like fresh firm soil embedded with seeds of thought my woman is mine my mind is her wisdom her foundation is my knowledge the creatures were given command to create under the watchful eye of God by genesis creation law and moses the evening is a day morning was made old day the old day is yesterday this was hidden from us evenings and mornings we are on punishment if we do not are dutywhich is for the mass earth to witness the devil ran off his harm is done the sun moon and stars have become soldiers with God and his angels the earthquakes above thunder and lightning earthquakes below the earth crust put God not to the test in emotion we were created love is the fulfillment of all the Gods who come in the person we will never love the devil yet we make the pin remind him of the sword the Gods and devils have not the same thoughts his way is not are way as far from heaven to earth are ways are higher then devils spirit is ones nature ones characteristics ones attractions ones actions by the spirit comes the attribute the name is king a crown to the head the devil was taken out of the black and brown germs the pairs of were come CaucAsian who would not exist with out the black and brown man atoms are of no space so in the stars that are dead yet we can still see there light by atoms in depth we are empty in a ocean of darkness we are the best part of life the children of God it is not good to be poor yet by wealth is how the rich control the mass with tricks and lies the mass know not so they kill 5% leaders for devils my God is not your God your God is not my God we think not the same I am 5% of 10 no blood sucker I suck breast milk we meet at the equality of 6 which is the limition of emotion some men cannot pass 6 and act like girls to pass from 6 to 7 is the ability to protect and maintain your 6

The utopia the horn of Africa

Ethiopia
Egypt

mecca
Palestine

Mathematics of the child is from conception nursing crawling walking early education and maturity to be born again is the seed in the womb and being immersed in the waters of baptism in order to be born again by the baptism of God the clot of the black women is the egg magnetic to the moon the moon was taken from earth by one of are black scientist who made the earth a lump of fire by one of the black scientist the God Shabazz was one of are scientist 50,000 years ago the idea of a God the father devil was known to Shabazz before this God was born who was to make a people grafted from the seed and egg of the original man of the original man to make a devil man yacub [Jacob] by his science it was known there would be no peace in this new people so the God shabazz thought to make the original man Asiatic in the west of Asia yet he could not make this man in the east so he went west and made a utopia at the horn of Africa a strong people lost and found in the far west we are the tribe of Shabazz original men and women made to run the devil off the planet God has come to are people in the person we must know and the wisdom of are words must be clear

1 knowledge is the foundation of all things in existence
2 wisdom is the manifesting of the knowledge one has collected
3 understanding is the mental picture projected through knowledge and wisdom which makes the 3rd eye a clear reality love and respect is the highest elevation of understanding which is the force so strong between man and women it cannot be broken in bond

A king is the founder of his kingdom so he must acknowledge the wisdom of his circumference so that he can enforce universal law truth is right right and exact at all times the original man is deity yet we can only pray for man the God we pray to is the first in ranks who gave order to be and it was this command to create was given the sun the sun of man the suns of man the stars
The original man is not a jew which is a chosen CaucAsian the CaucAsian is a gentile made by Yacob [Jacob] the father devil a God in the science of magnetics of the body to create the devil in flesh and bone grafted from the Allah body of Islam the devil has a beginning the devil has a end the sun moon and stars come not from the devil the devil is to end before the stars which are not diluted mixed or tampered with in any form there are 4 devil heads in the mind we must murder and run of are plane by thought the religious advice is of no force in
There religions yet we must free the mind off what was yesteryear to what is today here and now God is in the ranks number 1 the first he cannot rest until the devil is ran off of here if you are a man of understanding God is not to far off he is near or with you

The Creation is a work of art from God
The universe is a work of art from God
The Man is a work of art from God
The body is a work of art from God
The organs are a work of art from God

31

The 3rd eye is a work of art from God
The brain is a work of art from God
The heart is a work of art from God
The sex organs are a work of art from God
The religions are works of art from God
The Cultures are works of art from God
The animal is a work of art from God
The life we are given is a work of art from God
The karma from where came will return to is a work of art from God
The resurrection in are reproduction are return is a work of art from God

Colors are works of art from God
The things we read are works of art from God
The writings we write are works of art from God
The learning in education is a work of art from God
The poets spoken word is a work of art from God
The music of the drum in Africa is a work of art from God
The buddha
The hindu
The jew
The Christian
The muslim

Religion is no more then advice the culture of the God is the science and mathematics of the universe

The God
The king
Lord
Master

Peace is a universal greeting coming out of love a form of respect because if there was no love with out respect of others there could be no peace milky way clouds that move night and day the clouds move relentless devils have no part with God devils are not partners with God it must be clear God is the father of the righteous and will not sleep or rest until the devil is ran off the earth plane and by the angels pure and holy light the first of God was light from were those in the darkness were made to see out of black firm soil of creation the ink used to be written of are deeds in the light the mother book which is the page the pens are the black ink of the universe like trees are pens oceans are ink the trust in wealth in the heart is stupidity we as original people by trade were made slave to a deaf dumb and blind mental death and power we are lost and found in all ways acknowledge and recognizing God in are affairs God saw to it the angels with him were pure and holy good in all there nature the sky waters and land was taken from original men in the days of trade we were given the bird of sky fish from the waters the cow from the land and they keep the rest what was once ours as black people the CaucAsian is among us throughout the earth yet he is the leader of the poor part of land which is the west Abraham was put to study the sun of man the suns of the stars for guidance the birth of the tribe of Shabazz is written in the sky the Jesus was written in the sky the devils end is written in the sky we must count on the stars in there signs season and dates the sun rise is the harvest of the day the sun set is the sow of the sun as a seed is embedded in firm black soil were we see the stars that have been produced by the angels of God the moon is black it borrows its light from sun the stars light is 7 from were we come some stars are dead yet they still appear to the eye this is a sign of resurrection the body has germs as we grow we rot rust and mould the flesh has 7 heavens are soul is the aura of the [soul] that radiate in the clothing we wear the 7 heaven is outside the body by what we wear as are covering that is the aura we dress are soul not in the likeness of the flesh or the mind yet in the soul are clothing reflect the soul the Gods will transform the questions to answers the primal waters of the original black woman gave birth to the reptilian she is that old

We must exercise the brain being a muscle

We must exercise the tongue by the wisdom of your word eat by the tongue muscle

We must exercise the heart being a muscle

We must exercise the gut feeling of the intestines being a muscle

We must exercise the sex organ being muscles

The spirit will enter and you will speak the spirit will be within you and wake you to the judgement of all judges Abraham was said to call on the sun moon and stars by there true nature being in the name of God there nature is to dress the universe to serve man and not to be served the almighty is the only entity worthy or prayer adam came from the clay of planets he has always had the sun a clay on fire the clay and fire of the universe they have not been diluted mixed or tampered with in any form adam was holy and pure as the clay and fire of the universe until he ate with satan and was made low taken adam from his place in the heavens the original man is older then the clay of planets and the fire of are sun which is clay on fire the original man is older then clays and fires of the universe the stars and planets Islam must be established as a way of life a language not which was kept by Yacob [Jacob] the language of bable [Babylon] is not Arabic or Hebrew a God language of science and mathematics of the universe adam came from original people who were spirit yet adam must return the dust back to the earth the family tree of life was original men and women the family tree of knowledge of good and evil was of fallen angels adam ate with them when adam ate with the fallen those who interbreed are unlawful yet it is lawful for in laws to pass seed his seed was made germ a vice in the earth there is no peace between the devils and Gods the devils of adam seed the followers of the father devil Yacob [Jacob] a black scientists was sent to the caves and hills of Europe from the holy city of mecca at the horn of Africa the utopia we are free to prove God as we see him which must be made clear in the understanding between two three or four mathematics are a add day and night by the sun moon and seasons the sun moon and stars have there seasons on the earth given place times years and days they have a beginning yet they cannot end the physical will die in to the spirit this is the first resurrection physical wealth is the root to all evil most given wealth are blind to there possessions cannot see the value of there worth to God is better then what they hold God is the true wealth that comes with a return physical and spiritual wealth are both lawful yet we need both here and now with spiritual wealth in Islam comes physical wealth in Islam trade and traffic is the exchange of goods which can be those we feed or how we eat mental or physical we build the dumb and destroy the devil the lessons must be made clear this is the best food to be given the poor we feed the mind we feed the poor so they can eat by there own hand the mathematics make income because they make sense Islam is a language of numbers and words that show and prove mans relation to his universe which is God and light the black man is a valuable resource same as the earth value women has a wide 3rd eye she must witness the labor of the child she is most necessary her only limit is the speed of her God she can be secondary to her God by his travels he is her wisdom she is the knowledge of her God her God can be nesscessary or her God can be secondary by the speed of her God in his travel by the moves he makes the original man was the first to read the stars from earth and count seasons times years before they come emotions of the earth is weather the birth pangs through out the universe leaves paths of destruction through out the heavens and earth supernovas thunder and lightning and earthquakes with out earth weather which is womens emotion to create the child with out emotion no plant would grow no child could be born the womans 3rd eye is wide open to wise words if she is attracted she will question your motives the brain heart and gut are the trinity of enlightenment a king is the wisdom of his kingdom his queen is most near her God as he is most near her earth body the jew is Hebrew in there beginning the Asiatic black man is the Hebrew of the horn of Africa the utopia the Ethiopian church found by menilek 1 the son of Solomon the root of Judah the jew

Osiris [aser] was the first to be circumcised he was the fertility of the pharaoh the mark of cain 7 times and his son 77 times the seed of cain by noah seed was protected by 777 the father of noah lived 777 years when he passed death by the Gods Elohim Abraham was humbled with the surname ham of were comes the 7 trumpets the call to prayer shem the 7 bowls the mark of the beast is dis-ease of man and earth many are called yet the Hebrew is chosen before the name jew which is broken from the word Judah a black people the gentile is the CaucAsian the black is first to be chosen it all started with black the chosen CaucAsian is the jew a gentile which is chosen yet black is most old we are original to all people

Black
Brown
Red
Yellow
White

We are the lost Hebrew from the horn of Africa the utopia we are a lost to israel in are legacy are history has been stolen rewritten and taken out its text there is one God in the ranks the 1st yet there are to many ranks to count there are ranks all thought out existence who gives beginnings with no end is God yet what is not diluted mixed or tampered with in any form the 9 balls the planets and the ball of fire the sun will live yet the life of the devil will rot and rust the black man is to old to die he is most dominant black we come from black and black is from were we go black is from were we came black is most dominant throughout the universe the seed in rich black firm soil the black womb the egg the night sky the moon the sun of man to his solar system the suns of man to there universe the jew has no matriarch we who have a mother father and child which is one of the 1st of the trinity's jews have no matriarch they know not a queen they know not the God son muslims and muslim sons know not a queen they no not the trinity yet God can do as he will the Christian has the father son and spirit as a black man the trinity is known as knowledge wisdom understanding sun moon star man woman child 333=9 born 360 knowledge the circle is a-alike others who are 120 to b-alike others who are like you 120 to c- alike others who are 120 this circle is 360 you are 360 with your conrads a- b- c and on to divinity

Are culture is

Are nations
Are races
Are communities
Are familys
Are religions
Are circumstance
Are diameter
Are circle
Are foundation

I God all cultures they are all universal in advice culture is the cow the milk of the mother the queen bee the mother of royal jelly we all have germs that rot and rust are body the resurrection will be are right actions so we can have freedom out of the way are wrongs the right will give light to the positive the negative is not heavy enough to stop positive karma the black devil is the worst devil satan is not his universe not his character not his nature he is the supreme being the sole controller of his soul we control the soul it does not control us we lead it wrong or right in the battlefield of the mind with out stop with out limit which is a constant war devils cannot be taught Islam by faith or the sword and cannot be reformed we must murder there devilish ways in us off the earth plane israel is a tribe which became followers of Jacob by force in war and taken captive to the horn of Africa the utopia the tribe of Yacob [Jacob] became israel who are the holy Hebrews by Ethiopia Israelites are black man and woman the lies of Yacob were given moses the first set of laws were broken over the calf which was yellow in color not gold yet the calf was given birth this calf was worthy of life yet moses gave it as a offering to the priest there are many Gods to pray for yet there is one in the ranks the1st God is worthy of prayer and these two laws cannot be broken to who the word came were Gods and given to the Gods was the almighty who stands alone the oppressor

Knows not why the earth is causing him so much pain when he knows his falsehood as we know are truth God is most worthy to take on a son the Jesus called on God as a son to a father

The presence of God is in you
The person of God is in you
The divinity of God is in you
The science of God is in you
The mathematics of God is in you

Wealth is needed yet it is a poison root of evil yet spiritual wealth is to be held high by which comes physical wealth in its proper order the divine over material we must know are own reflect and make things clear God is everything and every thing is God from knowledge of self word is bond by the wisdom of your words which must be spoken clear to be made clear God is hidden in the black man yet we are hidden from no one we are seen and heard everywhere all through out the universe is black in shade the light in us is God him self there is darkness were there is no light stars are sands of the universe we are the foot steps to follow because we fulfill the books as God is are guide we watch from behind as we step on after knowing the right path set by God the moon is black the sun gives it light for its equality black magic in the bush of Africa

Superstition
Trance
Spells
Visions of wisdom dreams
Mediums

Wizards
Sorcery
Charms
Yin yang
Karma
Chi
Zin
Feng shui

The bush of moses on fire with out burn a people of no blame pharaohs made slave of his own people israel was caught in the middle moses had a hard time with the egyptions among him moses was for the I am that I am the one and only almighty those who were with him looked to the pharaoh moses was seen as less of a man yet he came to the fatherhood motherhood brotherhood sisterhood manhood womanhood yet he came to a dead people lost in knowledge of them self the pharaohs were men of all sciences of all existence by law israel was not slave to Egypt as the harm of black tribes in the wilderness who came out of Egypt to west Africa from the horn of Africa the utopia those who left with moses lost the law of moses for a law of a Babylonian influence Judah became jew the lost tribe of israel are sands throughout the earth most Israelites became Christians in the name of the Jesus some became muslims in the name of prophet Muhammad the body is the household of God he is the light of the body we are the temple of God we share the light of God we all have a portion of God him self all seen and heard is evidence of God alone God is the pillar that holds all existence in place physical bodys hold pillars in place the star of the body is the soul the pregnant womans womb is a temple the harlots womb is a tomb of were came the father devil Jacob who his followers created the grafted snake The CaucAsian the gentile by which came the chosen jews a gentile community of CaucAsians who are not from israel being a black and brown nation the jew has never been in enslaved john 8:44 the Hebrew from were came israel the Hebrew by moses was placed in bondage exodus 21:2

1 conception
2 image
3perception
4 rights
5 law
6 equivalent
7 nourishment
8 shrine
9 temple
10 attraction
11 unity
12 bliss

Earth is to reflect her man his understanding is on its way when she knows him as her God by the almighty in place of her man a God him self the child is born love hell or right the body is not the flesh the body is the spirit held in flesh we must give the flesh back to the earth and the body is set free to travel the act of creation was a commanded those who know not its science will be rejected ones way of life is the law of creation being divine math we reap and sow from the universe there is no force in religion we must give religion careful thought over a period of time religion is advice [add-a-vice] it is sincere yet its doubts can only detour you from the character in you being God the sole controller of the sun moon star sky ocean land fish birds beast man woman child leave them who doubt what is seen and heard everywhere God is the artist of all we see man was given command to be fruitful become many fill the earth subdue and subject the primal waters and the flying creatures of heaven be sure of God and give not way or be moved God has given advice by religion neglect it not it is sincere yet Islam is not a religion it is the science of all math beginning with no end came the sun moon and stars by science that cannot be broken diluted mixed or tampered with in any form these planets are holy are said nation has no birth it has not begun it cannot end are nation is in suspended animation and cannot be made other then old are nation will be here forever we are eternal are spirit body is immortal we are older then the children of God we as original man and women are the fathers and mothers of the children of God they were made in the image and likeness of are nation yet weak flesh will rot and rust the likeness of the Gods who are from the father devil the God Yacob [Jacob] will end in time Yacob did not live to see his grafted devil and his devil will not live to see complete rule God has come in the person to the black mass of the earth we are one God above below and under up down side to side back and forth everything everywhere seen and heard these things go on each and every day and every night in the body and mind sickness is poverty we all must die so take advantage and build life destroy death there is no deity with out the permission of God not with out ones own effort is one like or near God see God in your observation reflect and grow a clear understanding the best part of life is when we see him as are sole controller the giver of life to the dead or the taker from dead to life and make him clear to others by proofs of science and math of the universe which is true education we cannot have faith or belief which is not carefully thought out through time in order to prove God and man know God better then you know your self the 10% is the price we pay the blood sucker the slave makers of the poor who by 10% we give and are made naked and out of doors the 10% we give the blood sucker of the poor is made rich the mass population is given by the 10% faith in a spook and unseen God the earth is a woman gone mad she is a virgin yet all are not chosen and play the role of a harlot God is either with you or against you seek bread from God and this is to truly be feed adultery and fornication is to be unlawfully be naked except those of a binding agreement by time to clearly think out a binding heart a piece of flesh if sound between man and woman the agreement is sound if there is adultery or fornication the heart is corrupt the bodys of the man and woman are corrupt in heart be generous to your sisters and brothers and God will be generous to you by are hunger for more gold being the advice of religion when we were in tune with nature are gold was taken the gold of self knowledge the gold of are culture the gold of are truth the gold of are science the gold of are mathematics the gold of are universe the gold of are Gods a real devil is a germ of sin from the black and brown germ came a real devil being sinful by nature the pale horse is the white man death on his shoulders and hell given chaste this is why we are in the far west taken from are mother to build the devil up making the wilderness a concrete eden if pens were trees and oceans were ink man cannot increase the kingdom of God not one iota are mothers and fathers are elders we are brothers and sisters the children of are elders God to harm is oppression to follow those who harm is oppression to be forced to follow is worse then death God is the judge of all judgement thought is a mercy wisdom is a mercy the beat of the heart is a mercy the gut feeling of the intestines every joint of the body is a mercy the sun is a mercy the moon is a mercy science is a mercy from the advice of religion science is for big men religion is the food for children we must grow out of religion and science the universe those who are in tune with the universe are brothers the nation of the Gods and the earths are in tune with the nation of Islam we who are in tune oppress not each other we do not fool each other we lie not to each other we do not hold each other in restraint under force to follow a cause not yours God is most worthy to take a son out of respect the Jesus called on God as a father and Jesus role thought out the universe the sun of man the mans sun what makes man a God is his sun revelation 1:13-16 God is original black the Jesus was original brown the sun has made you black from the knowledge of self is by the wisdom of your words one is in paradise by the wisdom of your wife revelation 1:13-16 the ancient of days Daniel 7:9 revelation 12:5 the son of man produce what is your like and bring your proofs produce what you understand and bring your proofs the soul is hungry for life yet it is measured out in the goods and evils of its existence the soul is immortal because heaven and hell are forever there will be no end yet we must reflect the here and now before the soul travels on from its wrappings of flesh which must return back to the dust by the gravitation of the sun and moon on earth the moon is given to reflect the sun by the moon the sun shine with in the night sky the moon is most necessary to the waters of earth the doing of right is the only good the sun moon and stars knows no other then right and exact God is hidden in us yet we are hidden from no one the sun is a form of intellect are black messengers are guided by the sun and wise men are guided by messengers by who are inspired directly from God the pillars of creation are atoms unseen with the naked eye the atoms of solids have the most powerful bond they are stagnant in there bond a bond that cannot be broken the atom in liquids are weak in bond the atom of gas travel at a astounding speed atoms are the minute planets the blue print of the grand scheme of the sun moon and stars love your body or love your mind love the hell your body goes thought or love the right mind love hell or right a 3rd eye not used one is guilty the 3rd eye of the woman is vast and wide she travels by the speed of her man in the person of God if he is fast in his mind she is necessary and vital as water if he is slow in thought he is guilty for her care making her stagnant of a secondary the wisdom of man is the knowledge of his woman most necessary her understanding is made clear she is living to give life she is a creator a resurrector a reproducer she sees the child in the father to see the star in the sun woman is the moon the light among the sun and stars with out the sun she has no light she is black in equality with the night sky of stars her light is a touch of light from the sun a small and she is given life where there is light there is life the moon controls the waters of earth by the sun and moon the earth grows all its life by nature of are

1sun
2moon
3solar system
4milky way galaxy
5 universe
6 cosmos
7heavens

the atmosphere we all share clouds fog vapers mist air smoke from fire cold air of the lungs we see the spirit clear air is the nourishment as milk from a mothers breast as milk from the cow the vegetables we eat keeps us alive by the air they provide also keeps us alive the earth growth is vital we have no atmosphere with out air and water what we can hardly detect must rise cold or warm currents that distills back to the earth from the atmosphere to become rain hail snow thunder and lighting and earthquakes all this is done by experimenting on high exposions to perfect creation which is still a work in progress all in the atmosphere is caused by the sun which is the sun of man God in the person is here it is are turn back to the sound piece of flesh which is a sound heart circumcised of the flesh and made spirit we cannot live with out the earth atmosphere or growth we cannot live with out air and water air the earth growth or the moon effect on earth water we cannot live with out vegetation no air no life no vegetation in diet one is dis-ease one is sick life is the soul on fire the soul is the burning bush on fire yet not consumed yet it was of smoke which is the proof of the soul made black skin by the sun a devil is made of fire with out spirit the sun does not make him black he has a spirit of fire with no soul science is were we are at are most best over all advice of all religion science makes things clear science is the subtract in the eye a lesson in the eye is a teaching the interpretation of the symbols seen by the eye are the emblems known best by science the Jesus is the eye ear and word of God slavery is trade of bodys yet because some are not free in mind we are still at work for the mastery of are art and work the devil took physical bodys yet he cannot take the mind now a days because we are now awake yet some are conditioned to follow when we are called to fufill others to reflect and refine and become God in the person of man is your truth growth and development is law the jew is not a semitic he is a chosen gentile of the white race who are also gentile the horn of Africa is the utopia which includes Palestine and mecca we are the breathe of life known for the wisdom of are words the sun is black in its core its black clay burns yellow are soul is yellow flesh made black by the sun we are made black and own the sun the sun of man the sun is are wake up the war of Jacob with the so called angel was the tribe of israel the original semites Jacob took the name for his people the Israelites were grafted in to the followers of Jacob we are the children of israel the Hebrew know thy self and know the God in you we breathe God those of the left hand have wants in this life here and now those on right has wants in this and the next the for most in faith want the next life in this life we are threat with poverty which is the cure to up root all evil and not to love poverty or money yet we must balance the poor in spirit the 5% and the rich in foolishment given them by the 10% ordinary mortals cannot go to a mental heaven which is mind the abode of

God the so called angels are Gods angels are light the soul is the light of the body angels are through out the existence the angels are pure light life of God the Gods being so called angels are the nation of Islam that has no beginning and cannot end the nation of Islam which is a nation of original men and woman Gods not angels are the nation of Islam food is the fuel of energy power force sparks by which the soul lives by the mind the serpent power of brain skull and backbone the holy spirit is given by God as given to adam and eve for comfort we as original people are in comfort we need not to be worried God is here in the person were there is light were there is life it is all set off like a bomb from darkness the heavens and earth are held together by atomic bonds God is worthy a son are comfort in the son is for the father the kings bread is are inheritance a cheerful giver has the giving of a loan to God and this giving to God and others God will return in full from his house of treasures the word is made verse of the Jesus he is said to have spoken his word is the book of life we find comfort in the chapters of the books the mother book of all books is God given man is not a so called angel or a so called God we know not the black man is God in the person of man chosen above all power by the only worthy the God who gave start to all this u-n-I verse i-n-u man verse his woman steel sharpens steel the part she plays is most necessary by the tools of science we see atoms which look familiar to are soler system are 9 planets and sun is ours chosen for just us are soler system is the kingdom of atomic waters are soler system is a paradise no matter what you may feel or doubt the creation was created perfect the Jesus came as a servant of God to serve man not to be served by man he came to do the fathers works and gave rest to the Gods as the Jesus took up the cause we follow which is the God in the person of the original man we follow the Jesus he is the direct line to God who is worthy a son to serve as one unknown in the kingdom of all the children of God the light of the mind is God the feet follow the sound eye of the mind is the light of God he is the lamp of a sound body the Jesus came to give us comfort in the body knowledge of self is the mind that observe God in the person God is reflected in the body from the mind the Jesus is the only one who has provided a way of escape predators are lead by satan his wisdom is poison he is serpent the most subtle he is the man of sin he has a shadow of followers in the mass who doubt as he does those who follow the devil are made to be prey to judges presidents prosecutors police officers parole probation lawyers central intelligence agency federal bureau of investigation politicians the military psychiatrist medical doctors nurses scientists of the western world we fight not there flesh and blood yet we fight these rulers and authorities and there cosmic powers over this present darkness against the spiritual forces of evil in heavenly places the mass vote to these beast who cannot do what is not in there nature the predator is a wild beast there victims are humble and sincere and easily lead in the wrong there is no peace between one and another beast the predators are blood suckers of there own kind there is no equity in what they teach the 5% nation is man in the person of God a intellect r the Egyptian pharaoh 10% predators and the 85% pray the Jesus was made a target which by his work which was that of deity he was to be pray yet the holy spirit took his place the same comfort given his mother when she was weak the spirit was willing to up rise the Jesus from the dead those who know Jesus are given the comfort in the body of christ the 85% who is made poor by the 10% who were made rich by the trade and traffic of resources and black bodys of Africa to make slaves is wrong those in are circle of human kind were the same who gave the promise of more gold we became pray to those in high places we gave are bodys with out being given young animals are meek and humble by nature yet like satan grew to be beautiful in the knowledge of light this gave him the temperament of the beast which is what the predator grows to be a beast the 5% is man in the person of God who lives in the wisdom of his kingdom the predator is the 10% devil the 85% mass are pray to the 10% devil to protect and maintain your household is to up keep your woman the right foods mental and body the child grows and develops part creation law is the child raised under the marriage bond teach the child to ask questions which is most important to the parent and the child will grow one looking for answers must ask questions the breath is life a willing spirit of no end the milk of the mothers breast is the life we suck from the earth atmosphere the body is weak it dies every day by the germ of sin the 85% do not know the original Gods of this world who are the nation of Islam those who have no birth record no beginning no end elders of the nation before those in the sun moon and stars tomorrow is being made now tomorrow is created today yesterday comes now something is made new day to day the reflection is on yesterday the past is always now it is what you remember which is key the day is born tomorrow now is born by the past time gone and the future time to come the day after today the future the day before today is the past we must remember the times past and its lessons the past is gone by time yet we must observe the past and reflect on the now in order to see the future when we observe the past comes with reflection now the future by now and the past comes a clear picture for your future planned out the black man is the root to his earth

Right view
Right purpose
Right speech
Right conduct
Right means
Right livelihood
Right effort
Right awareness
Right concentration
Right meditation

The God in the person of the black man is the judge of all judgement the will is what you are able doing are rights are law written in are creation we must observe the lessons of nature these are the only laws in line with the universe a open eye cannot be closed the eye is the light of the body a eye of light is a body of light there is no savior to come to those who know not themselves to know self is to know the savior understand the God in you which is the person of God in man we must be free to have a meeting of the minds which is key to are communication which is peace between 2 or more where God is found the freedom of a people is by the intellect of the people if the devil is with you God is among you the fallen angels must be judged as to why they are angels made Gods wanting to take the place of the original man the God in the person the Gods of all nations other then Islam are fallen

Angels they are of the worship along side each other as long as there is so many Gods we can never have peace the God in the person of man is the true and living God and his holy spirits pray for the original man because he was before the sun moon and star the first is are nation we go back with out number are sole is the fuel that burns forever we worship God alone we pray for all other and not to we are not Gods or angels we are God in the person satan gave the advice [add-vice] of religion that we were angels or Gods yet we are God in his person you are not him he is in you he can end with out perfecting the eye with a new heaven and earth

Man is a teacher yet with out the woman and her teaching from the womb man is not complete with out her care from day one the family grows broken the community is broken the nation is broken the world is broken by just one family missing the universe is affected by this one family

I
Mine
We
Us
Ours
Me
You

Woman is flesh and bone of God she was created to rival the beauty of satan his was a beauty of knowledge she was created beautiful to the eye light to the eye her sizes shapes curves are round and plump she was placed adove angels and Gods to the eye sight there is no animal beast or creation to compare the comfort she gives the eye of man 100% are called yet few are chosen even less are to be anointed the slave makers teach a invisible God wrote the koran and bible yet the 5% teach the Asiatic black man wrote a history from were the koran and bible came we were taken out of the synagogue by the devil

The devil appears to be the lion of Judah the devil claims to be a angel of light God is with the black people religion is a spell are souls hunger for more then what we were given in the in the western world yet we will be rewarded in full for are labor we who are wise and civilized fall not victim they who know not righteousness is always the right way regardless to whom or what the baptism of the God Allah is when the earth was placed out side the heaven with falling angels the heaven was split one side God and di holy spirits the earth satan and his host the earth was made perfect a paradise and was made a world of devils by the father devil Yacob [Jacob] you are what your father is the elders are the person of God in the mind body and person of man we are children of the elders the children of the Gods in the person of the elders who has always existed before there was a history written a history from were came the koran and bible

The elders are grand living or dead they are old to the creation the universe in the heaven we as children grow in wisdom the foundation of the child is the father by the mother wisdom is made step by step as the child grows early education is most important to the universe the stars are inline with the buildings and monuments when the devil was in caves on all fours raw meat and came moses who could not civilize the know world by his works by which were after Yacob [Jacob] who made the know world uncivilized by a real devil not spirit spook or ghost as we are taught the magnetic tools of are science used to calculate number and measure was used by Yacob and moses there is no relation between religion and science Yacob and moses come not from religion they were scientist who gave study to are history from were came the koran and bible Africa is a book the nile is are blood are root is the culture of the mother land the horn of Africa the utopia Africa is a lesson to the eye a open book the book of life is the mother of all books by African science came koran and bible we as lost people are little ones to oppress one of us is to harm us all it is better for the oppressor to be not born what is done to the original man is done to God woman is the helpmate of God given man she is the characteristics of God the divine waters where the God cee him self in a clear picture she is the God in reflection God cannot see him self with out her his power makes her grow in wisdom which by she increase the best part of life which is being a child star a sun after his father we were taken from the true God and to all people were given a message other then the agreement of one God and one people yet this is seldom taught yet word is bond and bond is life fallen angels men not spooks traveled thought out the known world with message of there Gods and angels when the true and living God is in the person of the black man those fallen made them self Gods of all they could refuse and came force the government has no justice we have no rights to land as a nation of people we have no freedoms the law is not sound the duty of a civilized person in God is to approve and accept the uncivilized it takes a unit of time to make a civil the culture of Islam must be thought out before you enter black and brown were made those of hue the black babys angels in the sky the so called heaven the babys of hue red yellow white Gods of good and evil yet they are said to be just yet the court of law is under the judgement of God the scales will tip to there best God willing the babys of the black woman was given birth control gave her children as offerings by law the black women was grafted and the devil is liken to her child the woman was made victim by the sword of law held by the followers of Yacob who was the father devil the son of Isaac Yacob [Jacob] had no compassion for his followers he broke the law of magnetics as a atom is broken the bond of the black and brown pull on each other was made weak in order to make a weak people who were grafted out to the weakest point the CaucAsian by them there will be no more messages other then the 5% nation of Islam the Gods and the earths this is Islam in these last days a culture of scientist and Mathematicians of the universe the person of God in black people are scientist we have experimented with the creation we have measured balanced counted numbered and weight with magnetic tools of science what is hardy detected by the eye the tools of science makes things clear on a minute scale on a grand scale we can and have seen most worlds there is no mystery as long as we add on we as the original were the first to measure and in line temples and monuments with the stars the writing on the walls and tablets of the Egypt and Mesopotamia is the history from were came koran and bible faith is hoped for yet the evidence of are reality is that of a people who's faith is not seen we in ancient times gave witness to what was borne to us was a perfectly given reward of a sun moon and star holy not diluted mixed or tampered with in any form by faith we perceive the system of things were placed here by the commandment of God to let there be a re -creation of earth after the war of satan host and the host of God lead by the angel Michael satan came to make a perfected earth paradise void and empty of life to let there be what we see as a people is now in existence by things that were made visible yet some still do not know hear speak or see the devil was given a time line a set date among us to trade yet there was no peace with the devil and us by the arab and his form of Islam the sword was made law the Asiatic black man the original black in mind not color we have the most delicate form of Islam under God we are a people that must be handled with care by those who wish to trade among us we cannot be made the fool in this day the sword is still law above the head of the devil when the devil is among us there are forces that keep the universe in order in fixed form and measured out mind your ways and actions by getting understanding that we as a people are limited in are freedom rights justice and law yet we are in a court of a higher law from the judge of all judgement he is the only justice are reward is with are God with him and only him God in the person is the only savings grace the soul was placed in the earth to make the earth live woman is secondary to God she is the helpmate of God given man she gives life the breath given man was the woman from the rib God gave man life from woman who is one of the first in ranks of God the helpmate of God and man the woman of wisdom satan was the high chief of all beast he is the worse of all predators the beast of moses was of no help for the souls of his followers the offerings of the beast cannot atone we find no help as cain was no help to God by able as satan was no help to man in the host of beast as satan was no help to woman with his advice [add-vice] of religion which at one time was the straight path were satan was to come to the children of God the physical dead is dead the mental dead is dead the physical cannot revive the mental has a chance to revive by the conception in the mentel through the mind made physical in the womb from a thought the soul gives light to the earth and men are born you are the throughs you conceive in the mind and a child is through up by man and woman bear child in the mind then the physical child is born the void and emptiness of earth was by the war of the angel Michael and satan over the child the earth was given a recreation by the command to let there be light and came the work of his word freedom is the right to be deity in the person the 100% of the population are called and not just the circle of the 5% these chosen are a small portion of the called

The attributes of woman

Wisdom
Wise words
Mothers
Sisters
Daughters
Auntie
Niece
Womb

Secondary light
The moon
Old earth
The Gods and the earths
Earth vegetation
Water
Emotions
Supreme wisdom
Mother of God
Queen of heaven
Daughters of the kingdom
Wife of God

To be is to exist forever to have no end we taste death and return to the mind by conception we cannot return the physical from dead yet in the mind a life can return with new clothing life is food we eat and are given fuel energy and force the God degree is from the tree of life eat the food of God and find truth the devil promised more gold God promised more then gold God promised him self among us the reward for are slave labor God will pay us in full the devil wish to keep us naked and out of doors lies are with the devil truth is with God a mind of no light is dead the light of God are true body in are temple the true body is the soul in the temple the temple is the body the soul is in its house the soul is protected by the physical body the true body is the soul in the body of the temple the sun was made for are interest for are livelihood it is a star the best part of what you see is the light of this star are sun is the best star we were not embraced as a people we were said to have no soul we were made pray for the savage we taught to civilize yet the devil cannot be reformed we lose no time searching for the mystery God a CaucAsian cannot be God he is last in line of the ranks they follow the father devil Yacob [Jacob] the germ [sin] which came by the fallen angels and the daughters of man the fallen angels are sick in atmosphere not their own in these last days and times the devil cannot play us a fool by the lies of the devil of more gold then found in the science of the African bush and the horn of Africa the utopia which includes Palestine and mecca God has the last word God is the judge of all judgement the law of God is perfect of no flaws blemish or defect as the sun moon and star move with no disorder other then what was commanded in there nature regardless of positive or negative God is holy as the sun moon and star grow in to a perfection holy in all there movements the 85% ask not why they know not them self or the God in and of self the 5% nation question all things under or above the sun we ask why in order to be right and exact in order to answer all things which must be known in order to qualify in the small circle of the God 's by experiments with explosives by the tools of science are ancestors re-created and re-recorded earth by thunderstorms hurricanes tornados earthquakes and floods by these storms the earth that was empty and void by the commandment to be and the earth was made better in its re-creation the koran and bible are a record of what came before these books the wrights on the walls the tablets which were here before books of any kind the mother of the books is mother nature a lesson to the eye are scientist of old have been at work on the universe from a time unknown the scientist of to day know not us or him self messages were given all nations yet the God's they gave were not the same in nature attributes or character by the messenger honorable Muhammad we were given God in the person of the black man from are work with God in creation by are slave labor we will be rewarded in full compensation for what we made in creation work in slave labor what we made we own what we own is ours no messenger ever came to prove God by science until God come for are person as master ford God come to Elijah not a spook or invisible unseen not heard Elijah made known we are true and living God' in the person of God the original man the best way of life is the way of life with rights freedoms and justice in a fair court of law by the judge of all judgement God has the last word only God can judge us we must teach are women to protect them self we must teach are woman to maintain them seif we must be light as a feather in the sky when she is wrong in decisions we must make time and examine who we marry this is a court of law she is a nurse all woman fit to give care are chosen the child is Godly at birth which is a birth right the wisdom of a child is a privilege the moon is the second sun from the view of earth the sun is a star the best part of the universe the circle 7 we are not here to be servents yet we were made slave by a serpent who grafted him self in to are seed the Jesus came to serve us in to all truth and not to be served as we were made to serve the devil Jesus was the last servant to serve the lost found God him self is coming with the Jesus yet we must be comfortable with the one God of all knowledge God is secondary to none most necessary to all we were placed in bondage for fools gold we were made a fool by a foolish hand across the face the devil is manufactured by time his time in making slaves is over time in his doctrine is over his making followers even of his own is over to re-create the earth from the paradise it once was after satan gave it a flip up side down the earth was re-created by the angels of God the [Elohim] if the mind is not comfortable with the soul we are weak in the germ [sin] the soul is the gut feeling felt in the solar plexus the mind is dead with out the gut feeling of the solar plexus being the burning of life which is by the food we eat where comes energy to the body the soul can be sold it cannot die it can only change shape and form the devil takes given soul as the devil satan took angels he will take souls that have been given the war has just began by righteousness karma makes dead of evil we live by karma life is a door from were we come to be righteousness and burn evil the soul is a infinite spark of the mind the mind learns from the gut feeling the gut feeling learns from the mind the mind channels the soul the mind is dead with out its spiritual characteristics with out the mind the soul is sleep with out the soul the mind is dead a dead mind we can wake from yet a soul given is sold to satan the soul is only of a next to God in immortality the physical was made in the image and likeness of the spirit the war is in are own mind the war is with the heights of the mind the flesh is weak the spirit is willing to war on the behalf of the flesh hell is a house for those who have sold them self to satan the soler system is a atom on a grand scale the atom is a soler system on a minute scale the divine existence is the spirit the physical was made divine after the spirit was made physical today and time the scientist are ignorant in this western world they now not

The study of the soul
The study of spirit
The study of life
The study of mind
The study of the brain
The study of man and ape
The study of temples and monuments
The study of death
The study of rebirth
The sperm is a minute skull and spine the skull and spine is a sperm on a grand scale the mind travels as sperm to its destination which is eternal life keep your soul keep your mind the mind is the light of the body sell the body sell the soul we do not walk in the ways of the earth life and death we walk in the way of heaven resurrection and rebirth the black man is a state mind to be black is to be God the devil is a man of color he is emotional the colors it took to make the devil makes him color man the black man is the shade of the universe the color white man is a product of colors

Black
Brown
Red
yellow

the white man is weak in flesh and mind he was taken out by tools of science by study of two metals as the white man was grafted out of the universal family each nation was made weak black being the most powerful yet the white man is a real devil the germ [sin] in flesh dormant in the black man given life by weak flesh bone and mind a man made out of colors maken the white man the color man with out the white man there would be no color this is why the Yacob grafted devil is the color man white not black the germ [sin]the devils sickness Jesus came to reform the devil all the prophets have tried to reform the devil and his [add-vice]the advice of religions which when we as black people were scientist and mathematicians of the universe the sun is God to the earth by the moon which is vital to the womans reproduction and the earths atmosphere in the re-creation of the earth

oceans
lakes
rivers
hills
mountains
islands
deserts

were made by storms of the weathers

the re-recording of the earth was made koran and bible which come from the writings on the walls and the tables of are ancestors it is not robbery equality with God we are co-equal or comparable to God by effort God must be thought out by the information we collect we must count measure and weigh the knowledge of God

true hip hop

puba brand Nubian
krs 1 boogie down production
nas the firm
sean price boot camp clik
master killer wu tang clan
tragedy Gaddafi 25
prodigy mobb deep
killa cam dipset
big l d.i.t.c.
guru gang star foudation
q-tip native tongues
big pun terror squad
tupac outlaws
biggie smalls junior mafia
busta rhymes filpmode squad
red man def squad
jadakiss d- block
big daddy kane juice crew
Erykah badu soulquarians
Beanie sigel state property
Queen Latifah flavor unit

a man is the root to his earth the earth is rich black soil that holds his roots in place build science destroy the advice of satan given to are queens God cannot be held in a court of law he is law it self he is the judge over all judgement he is last word over all decisions he is the last say he is heights above so- called angels the black man God the universe with science to old to doubt to not know self is to doubt God the sun is like a bird through the sky the moon is like ocaan tides and ebbs that control the black womans reproduction the milky way galaxy is like the cow a wet nurse for the children the best part of the milky way stars you shell know God when you know your self we are strangers in a land not ours the devil is not a mystery as they make God out to be the devil has a wish to out do the works of God all the massive libraries in the world these books are not as vast as the mother of all books if the trees were pen if ink the oceans libraries cannot add not one iota then the mother of all books the lesson of the eye what you see is what you get it go's around the beauty of the two physical eyes are a reflection of the 3rd eye if we do not see what is best we are guilty by the law of God as God's we must clean are self up as a example to the devils that come among us the sword is used to remind the devils that he will be mentally taken off the devil of the study is CaucAsians the jew is chosen of his CaucAsian brothers the CaucAsian is the gentile

those not under the study we cannot reform him so we waste no time searching for a snake that does not strike does not exist the only God we have found is the child star the best part of the universe is when we are young the milky way galaxy beyond the sun and moon the child is born Godly by nature wisdom is not a birthright you must grow into wisdom the child birth right is Godly man and women have the privilege of wisdom we are born Godly by the nature of who and what we pray for

The Gods
The queens
The sun
The moon
The stars
The waters
The foods
Oxygen
Air
Warmth
Fruits
Vegetables
Firm soil
Earth
The heart
The mind
The body
The children
The flesh
Self
Animals
Fathers
Mothers
Ancestors
The black nation

The universe
The creation
Livelihood
Nature
Life
Resurrection
Good karma
Man
Woman
Holiness
Holy of holies
The womb
We pray to God only we pray for all under the rules of his law there is no above God he is the upper most God cannot be brought low and prayed for he needs no instructions we cannot pray to Jesus yet we can pray for him we can pray for the comfort of God in Jesus

Abbreviations for g-o-d

God of dominion
God one door
God overcomes devils
God omega divine
God only dwelling
God of demiGods
God of days
God of divine
God of dust
God often destroys
God of deep
God of depth
God of diamonds
God over doubt
God of degrees
God over dumb
God over division
God over dilution
God of detection
God overcomes death
God of duty
God over determined
God of doctrine
God of decision

turn to God and remember all will return 360 degrees life is a complete circle we are a sick people made that way by unjust law the only cure is the God of self in the person of the black man the cure of are sickness will be fulfilled that which is not God will be cast out by the mind all things are possible wise words by the pin

came koran or bible from rock and metals came writings on the walls and tablets being before the pin before the history of the pin we were a people of prediction by measurements size characteristics add deduct count and magnetics the great fall was the Gods made angels and then came devils with the white Christ and white angels and white Gods the black woman are the mothers wifes and daughters of all Gods the helpmate of God given man is the woman God is one in and of all his angels are positive and negative the black man is the person of God at one time the God was erased for mother and child the black Madonna

the father was taken from the circle of 3 all happens in 3 with out the father there is no family with out the trinity with out the father the woman is a harlot the child is a bastard the mother and child was taken out of context the

woman and child was made spooky a teaching for the dead father not the living fathers God is the son of the holy spirit God is the child of the holy spirit the son of man with out the trinity there is no knowledge wisdom understanding sun moon star man women child 333[360] 9 after the father of the family was written out of the books the black Madonna and child were written out of the books the God spell put many to sleep in a white christ a minority will wake from sleep those who are of mental dead the scales of God are just the guilty are all who have ears and all who can see yet they walk the street dead they cannot say they have not heard to have a mind is to know God yet the dumb are easy lead wrong all are born of spirit all things will be taught we will fall under the cure of just right laws where we will find freedom all cures must be fulfilled turn to God and he will circle you as the plants circle the sun we circle the sun as God is the center of all he is liking to the sun yet God is on a grand scale a existence around the God reflect the fires of the sun with out the 9 plants there could not be 9 in the belly with out the sun there would be no moon there would be no earth produce as the earth vital water is women the earth was re -created God let it be and came light from formless and void darkness all spirit is born to physical the Egyptian neter is God the Egyptian neteru are the Gods in the nature of the God neter is the place of the pharaoh his crown is the God head of the neteru he came to serve the Gods this was his throne his blood was the nile the God hapi pharaoh was made [happy] when his people were made [happy] the pharaoh was a slave to the dictates of nature the neteru being one in God one God one existence under God we cannot count the praise God has given in the reward of life not dead matter yet the physical has a spiritual natures characteristics features symbols and qualities one Egyption prophecy the God thoth [thought] was the Gods would come out of Egypt with Thutmose 111[moses] and took palestine yet isreal was taken and Judah was split Sodom and Egypt jew and isreal and those returned to Egypt the horn of Africa the utopia from were they would make way to the bush of Africa and made slave bodys to become of a western school God yet we are in a righteous house here in the person of the black man yet there are jews who are doctors of law Christian science gurus hindu Ethiopian priest the devil has no deep root the wind will up root him he has no firm hold the devils gravitation in are vine will be broken off forever in the court of law under God is are freedom rights justice and peace God has the first and last say God is the judge of all judgement we are with God who is the grand scale of thought who is the father of conception in the mind before the sperm and egg God most attract man to his women who are mother and father of deity the child made Godly by creation of mother and father made through God all children are created on a grand scale when the child is thought about sex is the positive force yet sex must be thought out before sex or child the judgement of man is not right with God they keep no law devils are hard to pardon a life of struggle comes with reward of no end we know not a resurrection from dead to life yet the mind is at war with high places in are own mind we must study the dead with science the mind channels the soul what the mind builds the soul is given the soul is willing to fight the high places of the mind and bring them low where they become weak flesh yet the war is not in are flesh God has given the worm of satan who eats the flesh war is in are own mind yet the soul is willing to fight flesh and high places to wake the mind is to wake the soul which is willing to war with the high places of the mind the mind looks to the soul for power the life of those in high places are condemned and brought low by wise men the power of the soul in the flesh is atomic take a atom from the body and split it it becomes nuclear of a atomic force the body is weak yet it holds a force of all forces the light of God and his nation that reflects the light and dispels darkness the darkness had a beginning and it will end the God of are nation is the light we reflect are light is God of no beginning no end in are birth record are God cannot be measured

the star cannot be measured up to are birth

the sand cannot be measured up to are birth

rain cannot be measured up to are birth

grass cannot be measured up to are birth

the hair of the head cannot be measured up to are birth

the flesh is weak to the devil he is the worm of flesh he eats at us morning noon and night it is sin to die with out knowledge of self which is knowledge of God we cannot travel to far in God with out knowledge of self the circle around you is complete when you knowledge it a father is the center when he adds on each and every day each and every way to his circle the orininal sin is when are spirit bodys became physical bodys by are multiplying with devils we are a breath of fresh air from God the devil is fire not light God is light and are nation reflects that light fire has no order there are no rules or regulations to what they follow they have no God or rulers they chose to be free from God and the laws of rights freedom justice and equality in the courts of law yet God is the first and last say in all judgment to sale your soul is like a marriage and satan is the host a soul for sale is given to a fallen angel and the soul is made black as the fire burns it the devil is a form of light yet for him to be in the darkness he must take your soul make it sit in the darkness and you and satan will be dispelled this is a form of suicide and satan wants as many souls as he can have a given soul to the devil is a dead soul the devil is good at what he does he has one commandment given by God to take as many souls as he can the war over the black Madonna being the womb of a black universe clothed with the sun and the moon beneath her feet and on her head was a crown of 12 stars she is the original mother of these lights she was pure in blackness given child to devils and man the archangel Michael and satan went to war and the earth was made dark of light satans tail began to drag a third of her souls to earth born to the Madonna from the heaven to earth and made them devils in a dark earth yet God re- created earth for man with the light of God in darkness and a reflection of that light is are nation older then the womb of the Madonna that holds the light of the sun moon and stars the Madonna is black by the fact she is the mother of are nation she is all the black woman who gave birth to all nations yet are nation is older then all the nations who have been given life we are spirits older then the physical birth of devils and man we are spirits of no birth we do not live we exist we are the fathers and the sons of the holy spirit by the umbilical cord we are given the spirit the royal jelly the mothers give us the holy spirit the universe is atomic the stars and galaxies are explosions as sperm is the star to the fertilized spilt egg is a galaxy made parts by the explosions of the sperm and egg the physical body is atomic on its in side take a atom from a physical body and it becomes nuclear of enormous force are wise words bomb on the universe to be unbroken is to be close to God we have been broken long enough when you as a king find desire in wisdom comes your kingdom comes your queen by the laws of your kingdom wisdom will reign by knowledge forever wisdom is a profit a reward to kings God is the minister of the marriage bed those who fight the high places of the mind are powerful of soul the war is won before there was a war to win before there was a beginning so old you cannot count measure or weigh when the mind has many concerns the mind burdens the soul the record of existence came with the bomb of God and came are nation we can count back to the universe which is just as old yet God and are nation has no beginning no end the writings on the walls tablets monuments and temples record the universe the koran and bible record man and his relationship with God and the stars God is first in ranks are nation is the first nation in ranks are nation is older then the universe and God is older then are nation satan cannot remain unless you will him the truth is square it must be measured out counted and weighted out in the facts of your square is the truth that will set you free yet most people are looking to rule someone or a ruler to save them some look to the dead to see some day some look for the dead to return some day religion is weak and frail advice [add-vice] of satan given woman in the books science is God given to understand is to be at peace which is the culture of the black man the divine those who are just must be at peace make peace or give peace take not peace no one can be forced to make peace we have been promised in the books a square of holy land 40 acres and a mule God given by those who are guilty this square is are own by the say of God are reparations must come it is what God has made are own by are labor of no compensation we are free in the physical yet the flesh is weak the mind is the fight education is key we must train ourselves with works are trade and traffic is charity we must buy we must sale give and receive at a price we are educated to be poor most know not self knowledge or from were they come the devil will try to destroy all he can

schools
church
government
entertainment

institutions
tv networks
health care
pharmaceuticals
the rap industry
Hollywood
Military
Police force

When being judged look for mercy that is with God a mercy not found in the courts of man the mind is dead with out the soul the soul is sold by the mind the mother gives the child its soul by her mind her mind must spark by her man and she is then open to give child of her spirit lifeless matter is given power man has not these works faith with out works is dead man with out woman is dead we see a clear mirror of understanding in the stars by the reflecting wisdom of the moon by which we learn and observe the sun of self knowledge are flesh and blood is a universal black nation the original nation before all nations we are found in the deep bush not yet know found in some areas we are found in Europe the poor part of land not because we are welcomed the fact is this is the last day we are found everywhere they occupy the globe we are the original population the war is not with are own flesh and blood it is the high places of the mind we fight to take from rob and kill are own is a seed planted in the mind some mothers do not give children a chance in the womb by drug use abortion adoption poor health care abandonment to harm the child in any way that a man and women bring in to the world you forfeit your soul to devils who will eat like dogs these dogs are mankind a kind of man yet with a devil nature hell is here and now are physical flesh is proof of this hell by the pain we feel heaven is made here and now life cycles is what we go through there is no pain in the grave to be dead is to be lifeless yet we travel though space to find a body to host mentally the mind cannot feel physical pain this is a fact yet the physical body can be afflicted it is fact that the over all reality is that the pain felt in the body is felt in the brain the God head between mind and body the soul is a next to God immortal once created it will be here heaven or hell the soul cannot die it can only be taken or sold for a small price it is worth much more karma is the door from were we come to exist to know karma is to know we will always be here death is the door to rebirth self is the resurrector those in pain cannot live in peace we cannot be at peace in a physical fire the physical was made divine after the spirit was made physical the spirit made the body divine the foods of the fetus are more refined and purified and richer then the mothers portions of food the unbilical cords are lilies the placenta is bud if the lilies as a queen bee gives the royal jelly the honey the holy spirit is given the child the mother is the giver of holy spirit the royal jelly the fetus is lifeless until the holy spirit is mother given a royal jelly the cow is a wet nurse woman is her likeness by the breast woman is the mother of are milky way galaxy we were here before the genesis we as original people we are to old to record we will be here as revelations has no end those who have a beginning will have a end no one can count or measure are history it is more vast then

the stars of the sky
the sand of the seashore
the water of the cloud
the vegetation of earth
the hair of the head

the holy spirit has been given to every man born of a woman satan from the heaven of God was one of those who would not bow to the earth saying he was from the stars he was fire and earth was clay and mud and man was dust the stars were made from fire the earth was clay and mud satan would not bow to the earth satan was made for the stars man was made for earth yet he was sent down to earth and made it formless and void to be re-created once a week there is a sabbath in rembrance of the 7 the creation of the righteousness work of heaven and earth are week is 6 in right work and completion on the 7th day

1 Sunday is light
2 Monday is sky
3 Tuesday is the waters
4 Wednesday is sun moon stars
5 Thursday is foods of the water and sky
6 Friday is offerings of food that are lawful for those in the liken of God
7 Saturday is the completion day of sabbath

If the dumb had God they would have light that would dispel light thought the triple stage darkness which can never over come God his light leads to the eternal and darkness will die as the last say of God which is the revelations that has no end what is done in the darkness will come to God the judge of all judgement God is are light God is a man just like you and me man and woman Allah the body of Islam Allahs nation
God is the light before there was light are nation is the light of God given are nation to dispel triple stage darkness by the light we became light out of darkness God created light were there is darkness the light of astrology is prophecy are birth date is astro in the stars is a lesson of guidance look not only for your own interests yet look also to the interests of others as your own the lessons give birth day and night is proof of the resurrection there is all ways a new day the sin of flesh is he in her man and woman must not look to flesh of he in her by the physical eye which is the evil eye yet see with the 3rd eye which is spiritually attached to all truths the 3rd eye sees what is meant to be seen not what we want to see the land by reparations swore as more gold given of devils is are 40 acres this land is the first and last say of God this is the only peace that surpasses all understanding no peace no understanding we were swore land by the koran and bible a holy promise not israel not mecca are labor is in the west God will reward in full the comfort of the womb is happiness as the coming of are reparations we must appreciate the land of promise here

in the west earth is are mother as Allah's nation we want are own place of worship under her sun the acres owed is just as important as mecca to the arabs as israel to the jew God is light his reflection is life to the earth the new book the book of life as kings we must have a book of life as he see the scriptures he must wright his life by the interpretation of his life and his relationship to God and the universe by the scriptures and his place in the text of koran and bible where they both say read and write God is are light by are minds eye we see those who will not see are those with the evil eye being blue they are guilty unlike the blue eye the original eye is brown the proof we are most in melanin not a blue eye gentile CaucAsian or his chosen gentile caucAsian brother the jew angels are attracted to the pure and refined angels return in the bath in a state of impurity bad actions are written down and good deeds are not angels flee from the unclean God is the only explanation of what who why and when are soul belongs to God alone yet it can be sold for a small price the love of money for the soul is the root to all evil the soul is up for bargain to be hidden as a treasure or sold as a possession a king is given the crown by the queen and child God is a judge of mercy the sun is the eye of God the sun is the glory of God the sun rise is the victory over the dead some will rise with the light of the sun others will rise with the darkness of the moon they have no light of there own other then what was done of light yet they no not what they do God gives mercy to these who die not in the Jesus and grace to those who die in the Jesus those who are in the knowledge of self have grace those who have no light of there own are given mercy the foundation is observation prayer is good yet to reflect on what you see to show and prove your words is even better this day and time we are Jehovah's witness we must make things clear other then what appear yet is not real being nothing other then illusion as long as we are given glitter for gold for are labor there will be no peace a kings wisdom is his a kingdom of jewels not to be sold for a small price information is the facts by time made clear the brain controls the physical frame the number 7 is the symbol of the divine black man the God of are understanding the black and brown exist with the fallen souls of the red the yellow and white nations black and brown reflect God as Allah's nation there is no culture that has a nation as are nation Islam is the last revelation of no end the qnat is weak yet it cause disease in man that kills man dead the yellow calf a sign of the red cow the law of the yellow calf was broken for the law of the red cow a child of God must honor his ancestors they created us as they made monuments temples and the hieroglyphics for us to know we are a noble people from nobles man is not to set Gods those who are worth the birth right have life wisdom was a birth right yet it it is now a privilege Allah's nation is the birth right of life before there was a genesis of no beginning and revelation of no end we existed with God are nation is before the sun moon and star were made holy by are existence unmixed or tampered with in any form from are existence came holy planets not by any iota can they be changed zero is lifeless matter given life by knowledge a zero is a cycle in and out of existence we live because we were re-born from dead by he in her flesh zero is a cycle of life and death and add knowledge to live again in all circles of life the baptism of Allah is the conception of he in her she is where the child is held and born by labor man is given mercy from the punishment of labor woman is given grace by the reward of children the sex instinct for the opposite sex is a need like water and food man cannot live with out his woman she cannot live with out him my knowledge is hers her wisdom is mine her weakness is her mans strength her strength is her mans weakness man is the root woman is the earth that holds the roots in place the past is the information gathered that you remember the present is the now the future to come is shaped by the past and present now builds the future prophecy which is said to come in the future by past events seen before they come by past events manifest now predicted in the past knowledge that is now fulfilled the prophecies of yesterday are in line with today made tomorrow the best of the people is the child the best of man is a child of learning the science of God made devils so the science of God can create a nation of Gods Abraham studied the sun moon and stars in the liken to his patriarch enoch who was taken in the person to the stars above to learn and walk with God as a angel Abraham gave us the knowledge as to the works of the heavens as enoch and its guidance not all of the jews Christians muslims hindu and those of the horn of African the utopia are not all guilty some are sincere some have a God of proofs a God not known by understanding is not the God to know the queen bee is for are healing it is royal jelly from the belly she was created to heal us with royal food honey womens placenta through the umbilical cord to the baby royal food the orinigal black man created nations and kingdoms by a small life germ nations became to race all populations were made of the black clot God is slow to anger he is not a man of war yet God becomes upset and his hand is hard by the chosen to war with all oppression israel is are land why give us back what has always been are land the devil will never fulfill his promises of peace in israel of the black man in the western hemisphere and are 40 acres the CaucAsian can save them self by peace in the middle east and reparations in the west the word of God can save yet we will never find peace with the devil we made trade on the continent of Ethiopia we were made slave the Egyption the horn of Asia made the Greek knowledgeable in knowledge being are gold by the devil you can live in fear or die unaware you have no choice in the matter with the devil the science of today is nothing new Greece gave us philosophy from which we gave them the truth is all around us in a complete circle pray for the Gods not to the Gods the names of wildlife is a early black science of domestication to control wildlife is mental not physical wildlife that cannot be restrained no help comes to those trying to tame them Egyptions were masters of this black science yet they know not the Gods of the adamites who needed help from God and became shepherds of the most humble of all wildlife flocks of sheep became wealth there was aggressive wildlife could not be of help of the wild the wealth of the nile jah is the letter of the law the true God is Allah the body of Islam the temple the temple of Islam is sound in heart and mind the soul is a witness of the minds instruction we are jewels by the wisdom of are words the black nation must come together as one a house divided against it self cannot stand religion is divided this is more of a problem to black people spirituality wealth is the nature of original people the us and we are the Elohim Gods who created men with immortal souls it by satan was told the tree from were he come of fallen angels could make the adamites immortal in body as he and his followers were yet the children came mortal in body yet God took not the soul from man out of mercy it was left immortal yet we are physical in body and hold the soul angels are immortal spirits the resurrection is not there's angels know as far as God gives they are limited to the law written in the creation what he has said be is here now and will move us on beforehand what has happen in the past is here now human life is sacred blood the child is in the blood of life made clot the child is blood which is life ministration of a woman the clot is made dead emotion is given by monthly labor pains every month she learns self the birth pains come with the month she is prepared ahead of time for the pain of labor she is a mother when her waters break some know they are devils some are with the devil and know not the 5% are the foremost of faith proofs reality facts truths by little faith we move mountains yet that is stagnation we must move from are little teaching of faith there is a time for all things time is are power by the days life we do the knowledge of all things day to day the koran and bible are not a mystery we are the people of these writings

Gods
God
Angels
Giants
Planets
Heaven
Earth
Hell
Prophets
Priest
Pharaohs
Kings
Queens
Wisdom
Light
Darkness
War
Peace
Love
Righteousness
Tree of life
Water
Wine
Milk
Honey
Temples
monuments
writings
scrolls
books

those who profess God are constant in elevation of mind by law the body must live by the mind the rules of the mind must build the body to completion of righteous work which starts in the seed of thought we must act right by God to exist is to be or end the hell you go through in order to come out right we must sit in the heavens now from light will come are heaven one day the are 40 acres is a concrete eden next it is a wilderness the next the earth is a desert the sun travels in the house of 12 sky chambers every 25.000 years the sun completes its circle though out the 12 sky houses comes a new record of seals in the scroll once the seal is open the scroll will be rolled up for new recordings to be fulfilled a day to God is a 1000 years the 24 hour of earth is 24.000 years to God we are a people who can withstand anything on earth we are the authority of heaven on earth the messiah has come this is the age of his light master ford is a Meccan he came by him self the Ethiopian did not come to us the hindu did not come the buddhist did no come the guru did not come Europe did not come God is the shepherd of the lamb and his flock adamite had no help with wildlife so they become shepherds

of the flock cain was a farmer by the seasons he know the stars as able knew the stars when herding the sheep his flock at night he would star gaze able was a shepherd of the most humble of wildlife as predators were no help to the adamite as they were to the pharaoh as wildlife was no help to the adamite able could read the stars cain reap and sown the earth God chose able to read the stars cain came to murder able by the fact able read the heavens and cain read the dust and the season when the seed was to be sown by cain the flock of sheep were scattered when the shepherd was hit by cain we have built the earth for over 6,000 years as Hebrew slaves a people of renowned by are righteous work we are due we must make a return to are status as kings and queens we are nation builders we are a people of universal order are mathematics found though out the universe science and the mathematics of the universe is not religion it is the culture of God every thing that we think in thought is necessary to wise words that which you cannot create in the mind cannot be created in the physical the earth is a spec of dust to the solar system that is gigantic to the earth to the universe is more vast then the solar system yet the soler system is gigantic the universe is more vast in weight the soler system is grand in size the days science knows no its size they know not the size of the universe the Gods are interchangeable with God we might be saying Gods and be talking about God we might be talking God and mean Gods the Jesus is interchangeable with God some time we are saying Jesus some time we are saying God at one time the universe was nothing God made it something after the first bomb came a star and from star to star came are sun we have been here before the stars by are records the stars are babies to the original mans existence the moon was taken from the earth by a God scientist and came pangea one land mass the adamite are a genesis people pangea was made 7 continents as one pangea was atlantis the garden of eden the flood of noah was trillions of years in the genesis not 7 days yet 77 trillions of years by 7 degrees until now the women broken of water is most necessary they create civilizations they create nations her waters are vital to life Allahs nation built up the nation of Islam we are leaders

Imam
Ministers
Rabbis
Medicine men
Gurus
Buddhist
Dali lamas
Monks
Hindus

The poor in knowledge cannot teach the poor we are poor righteous teachers the poor are the 85% poor in knowledge we are the poor righteous teachers of the 85% of the land mass of people on earth some women have a virgin minds and bodys some woman have harlot minds and bodys the woman is a temple her womb is the holy of holies man is the priest of her holies when he enters her he enters her soul and he gives part of his soul the soul is you you are the soul the soul is the true body is the body of the soul the temple of man is pregnant with soul the temple of woman has in her temple a soul a soul given to her soul from her man and a soul is born as a child she is the mother of a trinity in her soul her mans soul placed his soul in her soul and a new soul is born child the earth is blood the thick waters of man and wildlife are blue as the sky reflects the waters of earth we must take time out to pray it is good to pray yet meditation is just as good we must call the name of God when greeting others with peace we must reflect on God 24 hours 7 days a week day to day the days life knowledge the day of the 360 degrees of tomorrow in every days life tomorrow is born today the soul is God in flesh the soul is a eternal light in a flesh temple as a star in the sky which can only seen by day the soul will not be seen in the grave the bones are there as a testament we will be risen up if you held the soul up it will give God words of wisdom about you if you did not hold the soul up it will make a testimony against you there will be judgement by how your mind and body were used which has a effect on the soul some will be risen in the mental which is a physical resurrection God can raise up bones by his will the universe is still under construction the 7th day is the completion of are enslavement God does not sleep we do not sleep the devil on the 7th day will be put to rest the devil has gone to far to come back God is the big God the Jesus was a small God to teach the lost and found how to be a big God people the God tribe of shabbazz who went to the nile valley of the Meccan from east Asia he and his tribe left Asia to prove he could build up a big God people who can withstand any thing on earth or the heavens a man of no knowledge cannot build with the God man how knows not him self is subject to a civilization not his we must have understanding of the mathematics which are found in this life in self and in the universe mathematics are found in the order of space and time the birth of every new day we do to knowledge the days life the Jesus gave those in the grave of mental death eternal life we are the likeness of God the Jesus and self we have no need to look to the sky or follow a direction the land of zion is are 40 acres the God degree is mental and physical foods to eat these are the last days of the wrong foods the rules and the rulers over earth are upset they cannot control her the seals of the scrolls are now open the black child can now learn and study him self moses was a pharaoh he took scrolls from all the known world at this time the 613 laws were 60 books yet he broke that set of books and made ten commandments with 365 negitive commandments for each day of the year for the sun and its daily coming 200 chapters of the book of the dead coming forth by day and the 42 laws of maat and the star of David the up and down pyramids of 6 points of earth and the heavens above the moon and mars we have found monuments and temples this we know yet they know not the planets or there many moons 200+42+6 = 248 laws to keep the body in good health 365 laws are positive the sun divine a new word day to day the sun is not sorcery black magic black arts white magic witchery voodoo the sun is prophecy over prophets the sun is unique to every day he comes with clouds clear skys storms snow ice hail famine the sun also burns the earth the moon wets the earth when the sun burns we must knowledge the sun day to day the 613 laws are 60 books the 365 laws and the 248 law is 613 we must reflect 365 days a year the 248 laws are the 200 books of the dead coming forth by day the scrolls of Egypt the 42 laws of maat the daughter of the 365 sun degree the star of rasta is 6 points stars are found in 6 directions north south east west above below there are some in the scrolls who have studied the stars God gave us lessons by the star

Enoch studied the stars and walked with God as God gave instruction on the mechanics of the universe and how it works

Abraham was given wise words that the man women and child of his flock would be stars that would be enslaved in one of the worse mass migrations of a people

Joseph seen the sun moon and star bow down
The pharaoh the sun the wife of joseph the moon the 11 stars the pharaohs of the 17th dynasty were Hebrew and joseph and his people were at peace the 18th dynasty the Hebrew was byat war over how God was to be known king tut was the last king of the 18th dynasty moses was a pharaoh he placed his own into slave work as Jacob (Yacob) placed the loser of the war with the original israelites who be came 12 tribes with the sons of jacob Israelites that were born of a light skin were keep those born dark skin were murdered until the lighter as man could get the science was complete the devil is with man in fallen spirits yet the physical devil is 6 or 7 1000 years old

Moses studied the stars by which the people came out directed by clouds at day and stars
By night the 5 books of moses are not the books we now have moses took egyption science egyption wealth egyption royal woman and child moses came to the people yet they were not

civil with him we must not waste are time which is are power when you master time you master illusion lose no time it is on us to use wisely

King David studied the stars he was a shepherd after the heart of God he was a day dreamer he study the weather the sun the clouds the moon the stars the planets this is why Solomon was given wisdom not alone did Solomon rule his people his father David was a man of war God is a man of peace king David the son of king David the star seen by moses and in the wrights of the 5 books Menelik 1 was seen to come as ruler of the Ethiopian continent

The queen of sheba (Shabazz)the God tribe of Ethiopia studied the stars and seen her reflection in the house of Judah in her sky house

Persia arabia and Indians study the stars the 3 kings saw the child a small God to teach a people to be big Gods liken the father and his grand height the magi of gave a double crown the messiah which came now the wisdom of the word cannot be changed prophet Muhammad gave grace to the Jesus and mother we are in the age of the messiah he has come in the person the message of Islam cannot be broken

The magi gave the Jesus the frankincense to which would burn in his name in the temple by the high priest who entered the holy of holies in the name of the Jesus to give God a sweet scent in the name of the Jesus

The magi gave the Jesus myrrh in his works as a resurrector of the dead mental and physical coming forth by day the days life the days knowledge which are in the last day of the rule and rulers under satan we are in the age of the messiah the dead coming forth from a mental dead we are not free only in the mind

John the revelator the disciple of the Jesus was given revelations of the mechanics of the stars in the last days by which will come the war of armageddon in which we are liven here and now the last day of satan is written in the stars as the black man is to stand up as a people is also in the stars woman is the moon her sun is her man the earth gave birth to the moon as it now stands yet it was placed there by high science knowledge is the black man the wisdom of your words is reflected by the nature of your woman what you attract reflects you what we did not do to keep the law and what we did not follow in the law has us in the west freedom in the mind is a crown which is light of the body and feet a heaven now not in the sky yet God is above all time space energy enemies and evils

The prophet muhammad was given the study of the stars in the koran the heavens and earth are one and war will rip them apart the universe is self sun moon star man women child this is are truth in the heaven and earth all knowledge was on water to give life heaven and earth will not be one by war of Armageddon heaven and earth will be split in half the creation was made with portions written in the stars for us all is reward from God once we as a people stand up all creation is given size and shape by which comes are value the small measure is made grand the grand is made small the small of all creation is born before the grand scale yet we are a small people we will be made grand Gods between the heaven and earth is a war we are in the 6000 year end of the 6 the devils deal with us all the same in equity he will never civilize and act right and just by the earth we rise to the heavens the earth in are day has been shaken the reflection on the earthquake has come and is here proofs of the age of the messiah there will be no new message other then these of the Islam culture the science and mathematics of the universe the sun is a glory to man from the waters of the deep ocean which came waves made a fine mist the necked eye can hardly detect yet we as infinite souls are seen like the dome of earth reveals the clouds by the sun and dead souls by night cannot die yet they are evil forever made sick and look to healing are souls in the sky above the bones in the ground below the ocean deep is deeper even then the grave the 7 layers of the heaven are1 the ocean deep2 the ocean waves 3 tide and ebb4 the rising of the mist we breathe ascends and distilled and is seen 5 in the dome of the earth sky as clouds 6 by the moon we are blind to the night sky 7 the nearest heaven is the soler system the earth has 7 layers which makes it heaven now love hell or love right stars are the nearest to earth planets of are soler system are stars which are planets that hold the stars as pillers

the Hebrews were pharaohs in the 18th dynasty from were came moses (Thutmose 111) who lead the Hebrews out in there grief they wanted more from Ethiopians and a God they did not know the sun is a man the Hebrews wanted to leave so they were dealt with wisely for there works were not righteous moses gave a new scheme of the universe all in one and one in all the Hebrew inslaved by egypt lost all knowledge of him self and was given a God not in the formertimes known by father Abraham the God of moses was not the God of the fathers to Abraham God said enough to your search God is one in all the Elohim Gods or so called angels the kingdom of Egypt is not in the sky or over the sea it is a sound heart tablets that are circumcised of the pen as a record of what you have reaped in this life weighted on the scale with the heart and the feather pin which was pre-written before the heart tablet was written in this life the feather pin records what you did with your life what you wanted to do with your life what evils you did or did not do God gave us all a path to follow if we lose are way by reflection on God we can return and make way back to the God the universe is guide if you know the observation of the days life you know self and God in one equity which is not robbery the original man is a big God people from a big God the small God the Jesus came to make a big God people a man of no knowledge cannot build he is subject to a civilization no his own to civilize one must understanding of the mathematics of life self and the universe mathematics are found in the order of space and time the birth of every new day we do the knowledge of the days life we all have space around us as the sun has 9 planets these planet are dead stars the sun is a star yet the sun and stars are planets stars were grown from the beginning dead stars resurrect to new stars when the star burst it becomes new the cloud burst and the earth is made resurrect by the Jesus those in a mental grave were made free there is no prayer in the grave there is no knowledge in the grave there is no faith in the grave resurrect from a mental dead here and now todays life is the water of wisdom the sky house of aquarius the water can destroy or enlighten we must reflect on the floods the mass exodus of are brown people politicians wars and rumors of wars the mother of God is wisdom she gave God the firm black soil of darkness and God began to potter sons and daughters Gods who walked the earth with kings and pharaohs the Anunnaki the Elohim waked the earth they were not just in the outer worldly in space they walked with us and we are them on earth the mother wisdom of God was given man and women the wife of God is light by which the eye can see what the eye gives to the brain comes light to the mind the skull is the crown the spine is the tail the base of the spine is given life force as you reflect on the light of the eye your temple is given light up down from the mind down the eye is the power to distil which is to drop knowledge the child is given the power of time by which comes wisdom he or her must grow into wisdom yet infinite knowledge is all ways with the child do to knowledge the child is given wisdom when the child wake from the darkness of the womb comes the light of God being the wife of God she is are mother of wisdom are grand mother the mother of God wisdom by which all was created and made by wisdom all intellect is the wife of God the trinity of God is mothers wives and daughters the plan of God is to father all teach all civilize all God observers a day on a minute scale a day to man 24 hours to God this is 1000 years God is the judge of all judgement the soul is that of God its count is that of God the first and last is God said the mind says be and the soul be or end what it is being told the mind cannot end when it has infinite knowledge the soul cannot die it is immortal by God the spirit is your point of view your frame of mind God or devil the Gods gone live the devils will be broken in spirit the mind and soul makes you who you are the spirit is from the physical body and how you carry the body comes justice your body will be made innocent or guity yet the mind and soul will speak as to God for guidance those who know are no longer dry bones the soul has spoken lose no time it is time that is on are side time is a power for us to use wisely God was begotten by wisdom the mother he beget sons and daughters by the light he gave darkness he is the light his wife is the darkness which gave us shape and mold from the clay of the darkest night the only one that can match up with God is the mother of God being wisdom with out you there would be no God with out God there would be no you at its best a mind that reflects on God as you know him he knows you God was self created by the mother being his wisdom that gave birth to God thought out the universe God is being raised by experiments of high science on the universe as God works the creation each day is made complete for a new day the dead coming forth by day God is a child being light in a black womb light sits in the darkness light is a pillar of the universal black a universal people we are pillars of the heavens

God in self knowledge is the easy way to find God in the person find him and you are found there is no God with out you diamonds and gold is not the business of God he works supreme wisdom day to day of a brand new shine are born day is made new day to day you are you the numbers and letters which makes you who you are by todays life knowledge born self and nations are built women are spoils of war God is not a man of war women is of a infinite price by the children she produce woman are used in warfare those who use her in war are mighty to the powers of the devil the devil is the father of lies teaching rain hail snow come from God we cannot see yet these waters are seen and heard these are earth waters of the earth atmosphere the planets are dead stars planets have been proven thought out the universe the sun sits in the center of 9 dead stars the devil has a God we cannot see to the devils God is invisible they conceal the true God with religion yet God is a man of science the devil and his religion was made to make slaves all the devil is made to do is rob and live in luxury the devil was not made to serve he cannot save that is not his mission or his nature yet in the lessons of the devil we know the rules we walk with God the ruler of are steps is God the devil is on the path of religion were the said word of God they who walk with satan are guilty self is the only savior known one cannot be saved by others the temple of the body must be re- built life is the only reward with God the son is under are care when you are a son you are not a slave the slave is over the house until the son grows the slave is set free and the son takes the house the son is the messiah who gave the slave freedom the messiah to the lost and found black man God has come we are in the age of the messiah there will be no new message we are here now a culture of civilizers man should love his wife to death as the Jesus was willing to die for his flock we must complete the ciphers of life by time we master illusion which appear to be what it is not yet the Gods see what is clear to the eye we are planets as adam and eve were given a planet we by God will be given are own planets to become many and fill are planets subdue and subject are planets God is you and you are God the more space the more time God is to big to are inner space were satan was cast to the less space and the less time satan is now physical to a 24 hour day the Gods live by a 1000 year day time is are power time is are 1000 year day to day we add 1000 years to a1000 years this is how we do the knowledge most live not by time of God which is none other then time being utilize to waste time is devil walking after flesh is to be a student of the devil the mass 85% follow what they know not walking in the mind with God the student is the soul the gut feeling the burning of the soul the solar plexus all things work together for a good end by the say of God there is a good end with God by his work in us God is well able to bring a full reward if you abide by the wisdom of your words you are truly a student of the word of wisdom only a few speak wise words the mind cannot know what it cannot understand life is law the more we reflect on life the more we reap understanding listen carefully to what you hear the study of God is the study of mind and soul to renew the mind is a rebirth your genesis from a hell to a heaven which is written in the heart of the creation law the drive between man and woman is love how we understand God makes us who we are by the spirit of are nature God is are person it is a spiritual war information is the battle ground what we now know was lost we are now found in God the Jesus is the law of love yet the government does not have laws they cannot break how can they have love for you and me unless the law is made up of the rights to be free the powers that be will always lust to rob and live in luxury by oppression deep down in the devil there is no law he is made to break are circle of those around us to hate each other and love is lost as the law is lost the fulfillment of the love is are rights freedoms are social equality and are justice as to which we are big Gods in the liken to the big God and the small God the Jesus who made us big Gods the fathers the age of the messiah a new work will be made law we are a people who will witness fire floods storms war abortion murder crime hate and we will reflect on death and thank God for letting us see his work as he heals the earth with the sun those who know self and God and love God for revealing the days life of those not of the sickness of the mass 85% of the population on earth love was in are last of days past now it is a new day use the 3rd eye the brain to see things clearly the crown is the dome of the head which give a force to the spine the brain the 3rd eye spark the body by the crown and spine the life force is found in the sex organs the brain the crown by the spine distills and the light force ascends in the chakra glands of the body what goes up must come down a mind of light is a body of light the common man has not the things of God they are foolish to him the spirit man has a reward with God no eye ear or mind has seen the gifts God is willing to give the common man has no love for the spirit the common man does not understand the spirit the spirit man is God the foremost of faith have evidence by facts proofs truths and reality the wise are limited by the weakness and foolishness of the world peace is the seed of the understanding the fruit call on those things which are not as if they are in the name of God and what is will be the Jesus was a child of the holy spirit test you self by lowering your self in your measurement of self we are a big God people yet we must not act big over others are social equality should be a even path with all men yet it is not and devils measure us small this is why the Jesus came a small God to show and prove a big God of a people let your person be of love which is the fruit of understanding and the seeds of peace a master knows his season and when to plant as God planted the stars the creation came with a season by how you judge you are judged God is the judge of all judgement the God of all judges true justice is with God in the court of man there is no love for a defendant the judgement of God comes with love where we find are rights freedom justice and equality by which those who judge will be brought to light by the way they judge under God devils are condemned to the cloudburst of rain hail snow floods frozen ice and blizzards the devil has no place to rest in these days we have not known till this point that it is to meet the father in the flesh God is not a man of war the war of God is are war to fight some pick up war other then being oppressed these are those weak in flesh the spirit is willing to fight a mental we are the army of God with the full armor of God which is the full protection of God as mecca is to the arab as Jerusalem to the Hebrew as the ant hill to the ant as the bee hive to the honey bee as the spider to the web as the 40 acres are zion to the black man is just as important to are peace as a holy people who have found a holy land before your word is manifest you must know be for you move you must know where you are going you must know what it is you are saying and not just say it wealth is the root to all devilishment who makes the poor in mind state for a small price the poor are easy lead wrong once the soul is taken the poor are hard to lead right they are not poor as in wealth the poor as I what they know is poor the slave of this day and time is a slave to the dollar to be rich in the dollar is the goal of foolish men the fool has fallen to money and will do all they can legal or illegal for a piece of the devil pie it is not good to be poor yet fools sit not with the holy spirit of God and be comfortable in what they have which is God given to just have life is a reward from God which is the best of wealth the books and there laws will not leave or depart out of your mouth we will meditate on the law of creation in the book of supreme wisdom day light and night fall we observe and do according to all that is written by inspiration of wise men there is a modern day exodus of are brown people high fence and walls this is a yesterday made here and now on time 200,000,000 west Indians south Americas Mexicans and natives to the Americas chosen by the original man who are 17 ,ooo ,ooo original men who are Asiatic black man by the God tribe of Shabazz coming from Asia making it 19,000,ooo chosen men and women African by way of Asia answers come with study and questions are asked by those who no not the Jesus was said in the wisdom of his word he must go or the comforter cannot come he did go from the mind of the jew yet he left not in the recordings by the Meccan this is why which we know the son by the father if Islam did not record the Jesus we would not know the Jesus which is to know God by which we know self many call God yet few choose him many are called yet few are chosen they study the day and hour in the books about the last days yet God in the person is no longer a wait God is here and now

build all needs
destroy poverty

build to be well fed
destroy hunger

build what you must
destroy what you must

the best part of life is being a child as the child the Jesus the child is born in to sin not with sin

the child is blameless
the child is pure
the child is sinless
the child is upright
the child is unworldly

the Jesus is the child of God sinless as all children are in his like from conception a father must feed clothe and shelter his wife by the needs of the child the mother has a duty to provide for the child out of what you give her the wife is given no more then what you can afford what is given to the mother for the welfare of the children the killing of woman and children is forbidden the killing of a woman over lust or here evils is forbidden the killing of a child over poverty is forbidden children raised upright are a protection of there mothers and fathers honor the children are the future God is with us the son is God the wisdom of your word is God the wisdom of God is life in the Jesus is life and this life is light that gives life to man a light of body and feet comes with the crown of life the sun with healing in its wings along with Elijah to turn the fathers hearts to the sons and the sons hearts to the fathers before the grate day of war which is measured out of earth to come with this day and time the end by God 25,000 years not 24 hours God has measured the earth and time is up here and now there will be a new measurement which will be made the days life a new 25,000 years a new day the fathers and mothers are by the force of God made sons and daughters with all are hardships as a people will come ease as a people like day and night ease is coming hardships and ease will turn about each other the people of moses were bound to the law of the land of egypt the people of moses broke the law of the egyption creation law a school of many teachings so they were dealt with wisely as yacub

(Jacob) grafted israel moses grafted his people in to abortion of the son the law of Egypt was made vain by the people of moses now beast nations eat third world nations for resource we are eaten like animal flesh as the people of moses tried to atone by flesh when God is the original man in flesh the first to walk time and space the duty is on the black race the God tribe of Shabazz who made way to Meccan nile valley the horn of Africa and built it up by way of Asia the Meccan on the Ethiopian continent made one by sheba be merry make child eat drink and die the eye that can hardly detect the days life cannot understand because they can hardly see the Jesus is a virgin mind and virgin body we cannot be made a fool now a day God is closer to us then the blood of the body we are closer to God then the attributes they call him by throughout the globe when you hear the names of God you respond and when you hear your name you respond to the call the name is the nature you must study and live by the lessons of names numbers and letters must be built in order to know self when you are called call on God by any name u understand the jew must chose his CaucAsian brother or the original people his brother is a beast nation of sorcery lawlessness and hate and claim God the Jesus and holiness Allahs nation are Asiatic black men and woman we as a existence came with God as his reflection on what to create and we came up second in ranks satan was in are ranks yet he gave a light other then God to the mother are ranks are after God are ranks are far above satan and his fallen men so-called angels how can we be gentiles when the jew is a convert to Hebrew who were made slave by moses to the people of moses placed the Hebrew in bondage exodus 21:2 noah was black his sons were black the flood came to pangea one land mass atlantis made 7 continents the flood is of the old eden the Ethiopian land mass pangea garden made desert by curse on the adamites we are not a people of superstition or myth it is all real there is nothing the cannot be proven by the tools of science and mathematics before there was a CaucAsian on the globe we were building temples and monuments when we ran the devil to the caves and hills of Europe Solomon was built by wisdom Solomon was destroyed by wisdom we are temples as the temple of Solomon we were built up in body yet are temple was destroyed are story was taken after we were re-built the Jesus came to fulfill the destroyed temple after it was re-built in his body we became the body after the body of the Jesus the temple is the Jesus we are temples after the Jesus and God is are light the energy in motion is full of well lit emotion of God between man and woman love hell or love right the child must be born they are the future woman is most necessary to there production woman is vital to coming times of children those who have no understanding we must flee and waste no time with them and move on the life we live is for a reason which is for ehlightenment here and now so the child can grow tomorrow as a new sun appears to grow out of the night sky as if the earth gave it birth

The sun is heat
The sun is life
The sun is light
The sun is force
The sun is energy
The sun is fire
The sun is warmth
The sun is a ball of gas
The sun is the burning
The sun is the eye of God
The sun is the glory of God
The sun is the healing of God

The soler system is grand and great in size not known by todays science planets and there moons are not know the science of today they wish to walk mars yet there are many moons life could inhabit the outer space and time is of study yet the soler system which is the home of earth is not known the deep waters of earth are not known the more the science of today is studied the more is lost we are not alone if you can believe in angels why not do the days life on the Gods and other life throughout time and space the only reason we exist to learn God in the person be for there is death we must resurrect in the mind this life is to make your self clean to rule by the steps we walk God is the only explanation of what who why and when the father is king priest the daughter the queen of zion the son is the king by the wisdom of your words the tongue is a sword by which the soul and spirit are cut the wisdom of your words are life or death we must watch the tongue it can heal or condemn

The sun I am that I am
The star I am that I am
My woman I am that I am
What I see clearly is I am that I am
What I think up is I am that I am
A man I am that I am
The sun I am that I am
The star I am that I am
Light I am that I am
Fruit I am that I am
God I am that I am
My queen I am that I am

My child I am that I am
Wisdom I am that I am
Up right I am that I am
Righteous I am that I am

The holy land is not far away or across the sea the holy land is the 40 acres the black man and womans mount zion by time information is made clear time is are power by time we master illusions by time information becomes facts by the break down of what is true and what is false the Meccan came by him self

the jew did not come for us
the Christian did not come for us
the African did not come for us
the hindu did not come for us
the buddhist did not come for us
the Greek did not come for us
the roman catholic did not come for us

the Meccan came by him self we are in the astro house of aquarius we are on time in these last days the devil will end by todays math u now know it is are day are freedom is here we are blessed to be on time we are here to witness jah do his work jah is the letter of the law the astro house we are know living is a day of pure water given man woman and child in the form of wisdom we cannot be made the fool know a day because u now know we are the fathers of all civilizations re-given are place under the sun as the nation of Islam the Meccan found us and we will combine one to another the spiritual umbilical cord is connected to the light of the sun it is cut when the soul takes flight from the body and is re-born free in time and space the sun is the center of the 9 planets the solar plexus is the center of the body as the light of the sun travels through the body the brain and solar plexus are aware the sun is are life line are solar plexus is are the small sun the gut feeling the sun is the great light the lesser light is the solar plexus spiritual birth is when the cord is cut from the physical form the soul is God in the person the deity is found in the physical form divinity is found in the physical form making the form of man divine the meccan made are wisdom flow like water from the bowl to fill us with the waters of Islam after Islam is made clear by wise black men in the west can we move on to the east with are lessons only after the lost are found and made to know all black men woman child will champion the knowledge 0f the worlds we must as black people of the west we must be 100% of 5% as a nation the hour of are coming is now we are a second from reward we are on time by the minute of a particle we are on time it is are day and the end of the devil is also in line to end as we are coming to the jew the Jesus follows were a flock of the sick the old the young the blind the dead removed demons the works of the Jesus was deity to his flock of followers the works of the Jesus was seen as violation of the code of law yet he came to fulfill the law by teaching a spiritual meaning of the law the Jesus came to serve God not to be served by man were who know not God of todays life who is in the God in the person a child beyond its own understanding is a child of peace were there is no peace there is war were there is peace there is no war God is not a man of war a new born knows no war they are small Gods as the Jesus to God is a small God yet the Jesus is the maker of big Gods the Jesus in all his work the honor is given the father when cells become chaotic and when there is disorder of the cell comes sickness and disease death is illness of the cell health is wealth the healing is God in the person the mass 85% are dead in body and mind they have no knowledge of self to keep alive they know not the cell or brain sick body sick mind the 85 % mass are lost they have been found by the 5% to teach those in a poor condition of body and mind the first resurrection is in the mind of which some know not knowledge is infinite and cannot die it is energy of the brain to the body which the so deep it cannot be found by tools of science the second resurrection the devil will plead his case to no avail yet the mass 85% the scales will tip to the right side they will wake from sleep from the idol world of satan when the 85 wake satan and his devils will be taken off the 5% must be just to the 85% flock some know some will never know

God is on the mother plane
God is on the mother ship
God is on the wheel of Ezekiel
God is on the Egyption soler boat

time is the master of illusion it took time to find God as we looked out he was in we found God in the person time is are power we must observe are cipher what encircles us are creation is are space a circle of family and friends together as one in God most important the child is the future tomorrow has its own problems they must be taught to know from were they come and live by the days life dreams are not real u can wish for wealth yet the only reward is with God from the black and brown came
the rose gold Indian

the red bone
the light skin
yellow gold
white gold
pale

we want separation of are race from the pale throughout the earth geography they can run to the moon they can run to mars yet before you all are ran off give us what is are own and leave us alone grafting back will waste time which is are power time is given us master the illusion of time a 24 hour day to man is 1000 years to God the Gods have little faith yet we move mountains we are the foremost of faith

by facts as we understand them
by proofs as we understand them
by truths as we understand them
by a minute fraction the time of the devils rule will end as a twinkling of the eye as we move into a new day made old were the devil will be gone off are earth as a black people we were created right to do right and turn right as a sundial turn 360 by the right hand the nation of Islam is the first of all nations with no birth recording where a beginning is not found we are as old as time its self

build good habits
destroy bad habits

God is preparing for us what is in the eye yet it is not seen what is in the ear yet its is not heard what is coming in the mind we do not know what is being translated from God to the mind as a reward we cannot fathom what is waiting for us on the other side of slavey the prophecy is to be fruitful in the mind and and body and become many subdue and subject the earth to 100% nation of Islam throughout the globe as there is angelic life there is alien lifespan there is true and living Gods as there are false Gods there are upright men there are fallen men some have knowledge some have wisdom by knowledge of God we find are ability of are person is God what the eye see comes with a hidden reward lower your gaze you might see what is not yours hidden in your eye and you return to what was in your gaze which is not yours yet we will not see until the end of are time with out God knowledge is done we will rise up man and woman to a God complex the earth is a resurrector she gives life to the seed that cannot grow with out soil water and light the seed is dead with out the resurrection of earth a man cannot produce with out the resurrector being the black woman as the earth she resurrects fruit vegetation wildlife humans all from small seeds she is mother earth written in the book of life the devil will do what he was made to do which is lust then comes his envy his jealousy is then hate the devil will do what he was made to do what he does and that is evil we have found to know the devil is a lie the devil is who he is the nation of Islam is the first by time and space with no birth recordings yet we have insight on what and when by what we have the science and mathematics of the universe as we know it by numbers letters and the wise words of wisdom there is no beginning with are people the original man is older then the trillions of years of the star we are older then eternity older then infinity older then time its self we are to old to see the angel of death when we are bigger then death which is real to life as the mind heart and soler plexus and the kama sutra must be awakened and death will flee without knowledge of God and who and what he is the angel of death will harvest and place you with the army of satan a eternal death satan knows you do not those of knowledge have no end infinity is the foundation built up by what you learn and observe the energy built up in the mind and body cannot see the angel of death those in a poor condition of knowing are in a metal dead these kinds of people will be made the devil there is only a handful that will be take up in the mind to stand upright and sit the throne and break bread at the table with none other then the Gods with the wisdom that reflect the light of God with out knowledge you are worship of the devil with out knowledge add cipher one is cold to life frozen in zero dead flesh many were ice cold yet a few woke from the dead dry bones given life by infinite knowledge in are black essence in are black skin we are made wise by information the observation of life is key to life look listen learn the adamite was taken out of eden the white man taken out of the black essence of the black man amen is the completion of the text of bible be and it is done the koran is the opening the first text made to continue the bible text the mother is the black essence the black womb the black universe the God travels by black all wombs he enters all beings of life he is the light from were we have come a pure light the mothers the wives and the daughters are the black essence of a black God he is the first atom to be spilt and become the mind of the black mothers are essence God is the light of all life God has a mothers wives and daughters the light of God is the mind the woman is the body of Asia divine because she is black the clay earth the sons of God are from the stars as being life we come from other times and space as the snow we all are unlike in shape not one the same as a cloud burst with rain there are more drops of water then the count of stars the sons of God are from the stars and we are sleep to there existence the symbol of God is the sun we were created in the mind of God to use the mind we were created right to do right by God be fathers to all children brother to all woman she is your sister be a friend to all men the mother has no rank she is insight below and above she is below God above the sons of God she gave God the ability to blow like a bomb from her darkness in the woman was wisdom yet there was a son in the ranks the high chief satan among a mass of brothers satan is the father of a idol world he is loyal to this idol world he has the wildlife made idol did not work for the adamite so the mothers were made idol this is not the said word of God she was told not to eat with satan and his host yet she did as all angels have a mission the assignment of satan was to slow the coming righteousness of a dead people satan was given this work by God him self the angel Gabriel gave the koran the angel Michael when at war for God many of which are on the side of God will be keep from dead the angel of death is Azrael a angel in the liking of satan Azrael is a deal maker he takes the soul from body at a price there is a balance between the greater life the black essence and the lesser life the mind the mind must control the body the mind is God the sole controller the essence of life is the mind knowledge is with the 5% the bold in chest with so-called wisdom the 10% in the middle of are fight is the 85% who want understanding we are a holy people there is no God with out you there is no you without God the first in rank is God mothers have no rank they are the belly of the universe the womb from were came God the mothers wifes and daughters have no title they are all most necessary they are the word with God the word that was God with out her there is no God with out God there is no wisdom being the black woman who makes the babys she is the vital to the sons of God of which she is above yet she is below the God which she is below as we need God he needs the woman by wisdom who she is the helpmate of God in the creation the sons of God gave her eye in the form of a snake who first had envy of the adamite because he was the best make of God jealousy because he was low to the adamite the snake was made to be low that is what he was made to do that was his mission he was the high chief of all angels and Gods by his actions he was keep low when he was made as a angel of test and trial to man he does what God has made him to do the lust of the snake was for the woman he was called the most beautiful of all Gods and angels woman is the most beautiful of all creation this is from were came his hate he does not hate God he knows God as we know God he hates man who was created most beautiful existence of being the body was made from earth clay yet the hand that shaped us is God him self the body is divine to earth by the soul the earth mass is in a poor state we must not waste time with them be or born exist only to be here or born your person by growth and development mothers wives and daughters belong to God they are his the works of God are done in a strange way done in odd ways is his works woman was made in the mental image of God in the mind yet unseen or heard the man was made in the likeness of the sun the lion of Judah Solomon keep the mental image of God by building up the body of God by the temple the body of man the light of the temple is the soul to sheba the likeness of God to the flock of sheba the sun is the lion of Judah the seen symbol of God the creator in you is better then the creation the God in you ceated the creation policies of politicians and governments in high places of the illuminati and freemasons who will never give the truth of why we were put to work for no pay no wages the secret is how why where and for what the mahid is the person of God who came as God him self he came by him self to us the only people to come to us was those from the holy city of mecca the mahid came on time by a most minute second the devil is the evil of

abuse
cruelty
injustice
persecution
suffering
oppression

all ill treatment cause one to act out of character and become chosen not for the cause of the devil God is not a man of war the devil came to take peace from righteous man and woman for us to war with are mothers and fathers sisters and brothers the war is not are war yet some angels are predestined to war in there nature some men are chosen to fight those who harm the order of God as satan was made to do in his creation the highest fight is with self we war in the mind before the mind is upset there is no war the trouble of the mind becomes a flesh and blood fight satan is come to break are peace God is the peacemaker the mathematics are Islam and Islam is mathematics Islam is peace and the mathematics give peace by the days life the knowledge born day to day civilize the savage yesterday is from were we come now is todays life tomorrow has its own worries we are the kings and queens of yesteryear we taught as masters what is known which was built up to the knowledge of today which we now hold the devil tried to wright us off as not being the best creation with the ability to create a better add on over all the universe its self we are not wildlife we are angels and Gods we have five limbs as the five star the body produced in the womb is liken to God and his self expression by which he was self created light in the darkness as all life come of pairs God is a fully developed make light the positive coming life the negative darkness cannot overpower the light the womb cannot over power the seed we are God in are make a queen is the knowledge of God she gives body to the make of God a fully developed make of God how can angels and Gods have a make and not God we are liken to God in many ways God is all so liken to us in better ways which are not far from are grasp what we cannot find of Gods make we find in the Jesus the mind of the Jesus is mandatory to enter the kingdom the big God sent the small God the Jesus to make a big God people who are the lost and found God tribe of shabazz falsehood is done in the darkness truth is the light that darkness cannot overcome all wildlife has the geno to grow arm leg leg arm head life on earth began in the earth waters as fish that be came land creatures from land we took flight as the bird water land sky the bird is liken to the spirit that takes flight we are fish when we are in the womb we are tagpoles to the egg cell the fetus is a amphibian the geno of man that grows arm leg leg arm head we share with water land and sky creatures life is the Allah body of Islam

we share the body of Islam with fish wildlife and birds them being mother nature we are the authority of heaven on earth we are told to be fruitful and become many fill the earth with seed and subdue the seed and have subjection of the fish the ocean the bird the sky land and wildlife the knowledge of God is the knowledge of self Allah is the body of Islam arm leg leg arm head the way of life is the way of God the way of God is the science and mathematics of the universe God and man are undivided man cannot separate him self from God I shall be what is and what is I shall be I shall be and before I come to be it was spoken we cannot understand God when we know not self man knows not God by the fact he knows not him self a king is complete by his queen God is complete by his wisdom the wisdom of God are mothers wives and daughters she completes the kings kingdom with her wisdom the black man is the 5% of the earth population with the duty to civilization the cream of the planet earth is a man that draws the cream of the planet earth up to its fullest potential as fathers of civilization to Gods of the universe the 5% are the base of black and brown seed the 10% are a base of white seed thy to are called yet few will be chosen are seed and there seed are hostile from the womb we as original black and brown wish to protect are seed and not graft back to more a devil more weak and more wicked then the devil we know this is not one side some whites wish to protect there seed as we as black wish to protect are seed to graft back would change there future as it would change ares the future are the children they need mental slaves who by a mental handicap are physically inslaved we have no mind we can say is are own we are lead to the slaughter house as the 85% mass who know not what they do a wise women is a victory when won over the wise words of wisdom are very near in your mouth bread and butter the Jesus was the small God who came from the big God to make a big God people by the Jesus God became small so we could become big Gods you have the power to chose heaven or hell we put in the work to build nations as African people yet are story has been written out of history what was black was made white what was ares was made theirs what was black was made to be evil what was white was made beautiful the black woman was made ugly there are pairs thought out the universe that keep balance thought out the creation

man and woman
male and female
two physical eyes
two ears
two arms

two legs
sun and moon
heaven and earth
knowledge and wisdom
sky and land
predator and prey
fire and ice
hot and cold
penis and womb
sperm and ovaries
soil and rain
tide and ebb
light and darkness
king and queen
I and we
He and her
Father and mother

Woman is the helpmate to help man from his own self in order to call on God and were God is found self knowledge is found knowledge of God is knowledge of self woman is helpmate in this knowledge were we find God the image of God is what we live for it is are purpose of life to be as he would like us to be which is as a race it is not who wins it is him who has done his best a man of no knowledge cannot build he is subject to a civilization not his own we must father civilization by science and mathematic of life self and the universe mathematics are found in the order of space and time the birth of every new day we do the knowledge of the days life the pituitary chakra gland in the brain is connected to the two physical eyes were we bring light to vision were if we wish it be it is the pineal gland chakra in the brain is connected to the two physical ears the throat gland chakra is the music of wise words the heart gland chakra is the drum the soler plexus is the sun gland chakra greater light over the astro gland chakra the lower light is the physical gland chakra is the primal waters of space and time energy and matter are ancestors were jehovahs witnesses of the creation we live of the creatures we are told to subdue and subject they were believers brothers and sisters they come from dead stars yet they live in us and still live though us as some stars appear to be living yet they are dead they appear to live they are living as we came from the dust we must give back the dust from were we came we must see things for what they are yet there are times we must see what things appear to be to see what they truly are the dust we must give back to earth the soler system is to grand in size to large to the knowledge of the days scientist devils try to steal a hearing yet they find none God built his people to withstand anything in earth or the heavens God came from darkness which is mother her wisdom is the wife the sons are the atoms of a understanding universe the sons give the universe knowledge as the sun gives the moon light God is the baby of the mothers the daughters are queens of God the womb man a man who gives back the dust to earth to move on to a soul out of earth the woman the one who shapes the mud clay and clot she gives the soul she creates souls in the womb the man has not this power of 9[333] 360 triple stage of darkness complete in the womb yet incomplete must be born again in the mind 360+360=720 we are not born sinners we are born in to a hell not are own lose no time it is time that is on are side for us to use wisely the koran is proof of the bible the basic instructions before leaving earth the book of the dead [bible] coming forth by day the bible the only ones who can match up with God is the mothers and daughters and the sisters with out you there would be no God with out God there would be no you if there was no worship we would not have to pray God is a mind at its most the mind reflect and the power refinement= God wisdom is are mother raising children knowledge of self is the easy way to find God in the person is to find the God in self there is no God with out you the son in knowledge the daughters in wisdom time is are power waste no time on what does not exist be not zero in your circle you must add knowledge and from nothing something is made there is no were God does not exist we are zero with out the observation of knowledge as light to your planets grown from the beginning to be holy the place we created for us with out the stars we cannot exist with out us the stars cannot exist there are worlds through out creation dependent on each other some say the God is of one world some say the world are of pairs mother and child some say the world are of trinitys father son and holy spirit some say the four worlds are the heart beat of the drum some say the worlds are of five worlds of the wise words the music we speak six is a incomplete number it is the 3rd eye connected to the two physical ears if we hear it we want to have it and if we have heard about it we want it the seven is a complete number it is the crown what we choose to obsererve with the crown chakra gland is what we choose to see the crown chakra is key to the two physical eyes the crown chakra is a memory bank of what the two eyes see are knowledge is the creator of the stars the earth is among the stars we are the salt that preserves the mental and physical foods we must work to eat the wildlife is given a nature that comes with there foods are nature is not God degree if we eat the wrong foods the temple needs energy we cannot eat and have the force to live energy cannot die it can only change form this is why we die are energy changes to a new resurrect the mind and give the dust back exodus exit the dust we must keep the temple holy the right foods are energy that build the temple day to day by the knowledge of the days life we must build destroy and rebuild like a change of clothing rebuild your self the mental plane is the heart beat of the drum a state of mind there is more to are knowledge then heaven earth and hell the God of heaven and earth is not are God are God is a bigger God then the one we have been given the Jesus was a small God who came to serve the biggest God in order to create a big God people there is a higher being in man then the God we have been given if you cannot understand the wisdom of your words you cannot hold supreme wisdom God is to big for those who know not self there is no God for those who know not self the true and living man is God 24 hours is not a day a day is the rising of the sun to the setting of the sun 12 twilight hours+ 12 hours of sun light =24 hours the ways of God must be refined by man we cant be played a fool now a day we are in the age of the messiah Elijah has come the messiah we are in the last days and times of the devil and his rulse it is not a new day the messiah has always been here it took time to find him yet he is self knowledge and we waste no time other then self were God is not we are old before there was a existence we were here before there was a minute particle that begin to bomb on space and came time there are worlds beyond number the light of the temple is food it shines the body full of light the Ethiopian tribes know God before there was a God to know the age of the messiah is a world to bring us back to the old testament law the new testament law will be made old we must lean on understanding of God to understand self understand self is to understand God a master of time is a master of illusion utilize and see life is just a stop to test the energy we have built up to find the life we live is a resurrection of the mind only the foremost of faith will be given a past for the fact they orbit planets by truth facts proof righteousness and at a up right stance small faith moves mountains they move the earth the foremost of faith move the heavens God made us to own all that is his the original man is the Asiatic black man new nations of red yellow and white are old in us we have no birth record we as original people are the genesis the child is not a sinner he is not shaped in sin the child is born in to a sinful world sin is at are door men war over food it is the worst from of oppression a hungry man is the worst kind of man some will die for crumbs some will kill for crumbs the free choice to do evil is with the devil the free choice with God is all that is best choices are hard to make we can choose evil we can choose God we know what is evil we know what is righteous it is all in the gut heart and mind that which is below you is not you there is nothing righteous above you that cannot be reached we are the book [bible] of the dead coming forth by day we are not the 85% hard to lead or easily lead are nation has joined hands with God the mass are marked of the beast microchip in the head with out knowledge and know God not with out God we have no knowledge it is on the tablet of the heart were God is found knowledge is key to all openings satan is at the door yet he has no key his root does not match up with ours let him not into your kingdom God has given us the keys to open all doors actions cannot be made with out information to move them which is limited only by what you know there are no wise words with out the knowledge of the said word the black man is in exile from his own we will not be at peace until what we made we own by these facts alone what is ours is the earth civilization and the universe the divine way of life is the days life that comes by knowledge in order to walk with God u most know God and the closest thing to God is you the devil has enjoyed the fruit of the black mans back we are a treasure to the devil yet he has tomorrows worries we have measured out the earth and he will be ran off the sun moon and star are respected in there elder yet we are older there is glory in the sun moon and stars yet we are the authority of heaven on earth we own the earth we come from the stars the orbit of the atom are the roots of all things there is no space where there is not a atom there is no invisible are faith is foremost we see the unseen by what we see all cause has effect

sun light
rain cloud
food energy
intercourse child
sow reap
life death

fire water

as we have space to move we have time to utitize the universe is create a perfect energy cannot die matter cannot end the only fact is change we are free by the space we occupy by time we have measured out the earth and know it in its full energy is self preservation mental and physical foods needed to resurrect in mind and body there is no knowledge in the grave get it here and now matter cannot die it is held by atoms that hold atomic and nuclear powers when split a and taken out of a physical body of a man and it becomes nuclear this is the fact of are spiritual strength the atom is made to maintain the universe forever atoms are the foundation of all things the atom is a small solar systems we are in space on earth the authority of the universal on earth in solids the atoms move slow all most to a stop we cannot walk though solids the atoms of liquids move faster then solids yet they have a effect over time they have the power to cut though solids called soil erosion the atoms of gas move at a tremendous speed hard to detect by the naked eye yet we see its effect when it comes down as rain when the leaf is made to move the breeze we feel by wind the air we breathe by are oxygen the Islam of the 5% minority populace of earth is science and mathematics God was made far away by the enemie God in truth is closer then the fluids though out the flesh he is closer then all vice we are captains lieutenants and soldiers in the work of God do to knowledge the black man is God by greed in high places we were made inslave we did not know we were naked we were in a pure state of body and mind we are taught what is not real when we began to grow and develop in self we began to observe reflect and see the best part of life is from were we come before we were here in earth space before we were soul [light] before we were children of adam by noah before we were children of israel the Hebrew the stars were are children the genesis is the re-build out of nothing that can be recorded they say big bang to keep us in fear of that mystery God who is said to be dead coming or cant be seen with the physical eye we are a dead people it is God said one will come and we will see him clear as the day sun in all black man and woman the genesis is much old then the said word be who you will be and came the [light] the soul the highest of all light energy called the big bang yet we know though out the universe black scientist of old examine measure and bomb on the universe the creator is the genesis with in all though out the heart depths of the universe the only God is man God is no more then man the beginning of universal change new made old by theology of time by a fraction of a second we are on time the so-called end will be no more then change the original man is the Asiatic black man from Asia Africa is west Asia limit not the black man to Africa it is not the only home to black people God is the sun of man the supreme being black man from Asia by the knowledge of are ancestors spoke with wisdom by there wisdom which is in are geno D N A and chromosomes we gain power and strength are purpose in life is to give life as the God gave life to the space of the earth orbit is in space full of life are purpose is to take after God in all his grand and eat his foods mind and body there is no force in religion it has no power over science the black man is the worse snake in his core there is no true devil snake is not his nature the snake was mastered by the early egyptions the serpent power is the ability to suppress the serpent of body and mind the serpent power is the wisdom to limit the serpent in you the hindu was exiled for there form of Islam the buddha gave are the birth record of 35 ,000 years old which came the buddha a king was made of the most high in enlightenment a God with knowledge we all have this bliss which is happiness when we are happy with self the buddha is seen in the visions the Jesus was seen in the stars by all prophets 551 years ago came Christianity years after the Jesus came his teaching was not christianity the devils used his name to make dirty what the Jesus gave and call it religion the Jesus was free to serve justice reward or penalty were there is equality all will be served justice the Jesus came to serve justice giving and receiving is trade and traffic the knowledge of God is are business when we give it is a lone to who it is given the lone given others is a lone given God that will be given back as reparations we will be paid in full a righteous mans business is the work of God there is no belief faith or religion above the foremost of faith where the facts evidence proof and truths of where come the science and mathematics of the universe the foremost of faith are in intellect we must examine and experiment on what and who we worship and bomb on day fore day success and failure are of your own making there is no devil to blame

knowledge self

love self

respect self

we are the geno of God we and us are the same God is the genesis from were we come of who we are the adamites were a dark people taken out of the holy city of mecca for eating from the wrong family tree the tree of life is a family tree of some of the darkest black men and woman the adamites children from black + brown = red this red seed the adamite children begin making trouble telling lies they were told by the serpent of body and mind by this serpent power there will be no peace if not used righteously and overpower it the satan has come in body and mind to be suppressed we must work the mind and give the idol body back to the idol world of satan give the earth back her dust when we exit her clay we must strive for perfection by doing what is most best the children of the Israelite children the Hebrew and his child made slave for the image and likeness of the wildlife in worship up in till the fourth generation punishment for there error the Israelite will be hated yet the child of the Hebrew will find loyal love to the thousandth generation for being made a slave by there own people as the Jesus came to his own home yet his own people did not accept him before the Jesus was moses a God to the people of Egypt who were savage and given a heavy hand as the Jesus was deity of those with small faith and they were born not from blood or from a fleshly will or from man's will yet from God yet moses and Jesus were not on time Jesus in moses time was seen in the stars to arrive the Jesus came and the son of man to come to be taught as to be the al-Mahdi to come by the said word of the prophet Muhammad by theology of time the al-Mahdi has come in the person of God the 6000 year week is over there is nothing new all is the same day under the sun God came to the honorable Elijah Muhammad of where came the supreme wisdom when ford came we were given the holy city of mecca back as a home bread to eat clothing to wear a key to open doors where there is no peace there is a sword we are a people over all people yet we are in the worst condition of all people we are wooly hair sheep humble they are goat straight hair stubborn the yellow hair of the goat is recessive the blue eye is recessive the yellow hair blue eye man is recessive the Asiatic in late day Egypt placed the Ethiopian Hebrew of the black land of Kemet in bondage we were returned back to bondage in boats we are not built we are the builders the God make of the devil we built up to show forth are power that we are all wise and righteous knowing he could make a devil which is weak and wicked give the devil power with out falling victim to the devil's civilization other wise to show and prove that Allah is the God always has been and always will be the Gods will take the devil into hell measure and weight every days life what God does not do we will not do we were made in the image and likeness to become who God is

the command of God is the word

said
spoken
a voice
a talk
a speech
a statement
a utterance
a order
a expression
a comment
to declare
a discussion
to verbalize
a lecture
a suggestion
a dialogue
a language

God has the first and last say he is the said word as God spoke it was done and the creation was built by the word God commanded to be which is made in the wisdom of mothers wives and daughters the earth is the wisdom of the world to understand heaven is to be wise to the earth the devil took bodies on boats and made a slave the devil now the devil takes bodies behind bars and make a slave we must study the new to learn the old being who we are a people of the ancient day to day we give the day to God the days knowledge is given by God to man the God has a degree of wisdom by which the universe was created God is not unfair as man is unfair to man the father and mother produce deity respect those you love as you love your self

knowledge is infinite
wisdom is eternal
understanding is absolute

the true and living God is man the universal greeting is peace the names of God are salutations
when we call him there is a response as we respond to are own names

1 heaven to heaven
2 cosmos to cosmos
3 universe to universe
4 galaxy to galaxy
5 star to star
6 planet to planet
7 moon to moon
7 stages of creation

we come
we came
we are there
we are here

the genesis in the physical body is stronger and faster then the physical body we are created of self
by are own struggle and effort to exist we are the seeds in the fruit of Islam to observe is knowledge
were we examine weight and measure to cee clearly this is self mastery the wise words of the Jesus
are the words of God the wisdom of God the God we serve is love the Jesus came to serve not
to be served love serves us the Jesus came to give us love in a most precious time of need we as a
people were once dead yet one the naked eye could hardly detect as a savior to a black lost people
was master fard made one out of are mist the most honorable Elijah Muhammad who was made
to ascend higher and increasing with other mist being laborers who then distill back to the earth
like a cloudburst we are the tablet of the heart circumcised with the writings the heart is the last
book which is of old the heart is the mother book the truth of the heart is a book the heart is the
last book we are a old people of a old way Islam is the cultivation the caring for and tending to
the earth are mothers wives and daughters I sincerely love Allah s mathematics

the bee its hive its home
the spider and his web is its home
the ant and the ant hill its home
the sea life the ocean its home
the planet earth is are home

the earth grows the cream to fruit by the seasons of the womans months we have made a exodus from darkness to the light of life by knowledge life becomes more then infinite more then eternal more then absolute life is a mass exodus from there to here from here back in to the essence from were we have come clear the mind respect the universe the prayer is fulfilled before you make it prayer is potential energy what is potential has already happened yet it has no yet appeared until there is kinetic energy which is what will happen with out the potential energy of prayer kinetic energy is point zero it is a void that must be filled with prayer with out the potential energy of prayer kinetic energy is formless by prayer we are guided in faith that all will come to pass for the better by obedience to the truth

the jew

the Christian

the muslim

the hindu

egyptions

are not looking for a spirit dead and invisible they are looking for a man a physical man yet this man has come in the person of God to the lost found muslim in the west were fard the God in person gave the last messenger Elijah Muhammad ascension along with the laborers in the work of Islam we as laborers must work

the brain muscle

the heart muscle

the stomach muscle

the Gods and the earths

the cream of the planet earth the God of the universe with out a positive charge there would be no force to hold the physical in place the negative charge is why all things change we will always be here there will always be change

the sun is a proton[positive]

the rock planets are electrons [negative]

the gas giants are neutrons[neutral]

change is key to existence in order for us to grow is why are purpose is to live some forget the genesis in side is the atom a mental and physical body the positive cannot rule the negative cannot rule they are neutral when there is charge there is change for the better positive is most dominant and keeps us from a complete end

the communication of God to are people by music

drum
piano
harp
flute
guiter
violin
bass
harmonica
the communication of God to are people by nature

wind
rain
thunder
words
sound
song
noise
yell
shout

we walk with men yet not as men genesis is the awareness of the body in you what we have heard seen and know is not known by none other then the original man the lost found we are here before the self creation which is a self expression we let be more then one world all worlds are locked in to each other in one accord as one yet the masculine and feminine are the pairs of which are the locks of the universe that holds the creation together the Jesus by his own was told not to say child of God not to say children of God he said we are Gods the God can do as he likes as like we have found the trinity to be

1 knowledge sun man
2 wisdom earth woman
3 understanding star child

To be a child is the best part of life for some children to overstand the mother and father is key to the knowledge of the child who must be taught so they can grow and develop there own it is not what you want it is what you need and then supplement what you have with what you need then comes wants the mind is the energy in the dead matter of the physical body death is physical yet the mind and her energy cannot die it is the genesis of physical we were once dead a physical and did not know pass grave the energy called life is the genesis that cannot past on yet if you are mentally dead you are a physical dead get knowledge and live beyond the matters of this world and live as a magnet pulling on God attracting him by your actions Islam is the same yesterday today and tomorrow all comes out of the womb of Islam which is the elders nation before

Infinite
Eternal
Absolute

Are elders are the nation of Islam before space and time we were here there was a bomb the elders let be causing a chain reaction of which came cosmic slop a chaos which had to sit as forever so far back it has no recording so far back this chaos has no beginning or end it is forever perfecting this is were the sun moon and stars came a heaven of magnetism we are attracted to the elders this is where came the idea of one God came the elders have many names and they are all Gods yet the first in ranks is the most high we are attracted to the elders and there many names are called by are walk with the elders we attract the God in ranks the most high this knowledge is a larger form of wisdom this understanding is a larger form of knowledge and wisdom man and woman become child and child becomes man and woman sun moon and stars move day to day we must observe to be wise and cee the heart must be warm to the things of God this inspiration is to find the mind in order to wake you up the heart and mind must be in line we must know when to ascend to those in need or distill like a cloud burst and bomb on devils your mind is what makes you God the true mind is God Asiatic black that makes you the person who you are being God we all have are own facts we live by we all have what is true to us we all have a reality we go though we all have are own circle around us we have created the father devil gave a make to a devil of his weak and wicked brain who is here to give us hell yet hell or right truth or square 5% will come out the 5% are here we are divine the devil is divine yet the devil is divine in evil we are divine in God are truth is out of the ordinary we are not of this world of satan he is from are world we are
The great
The high
The upmost

The 5% rules and regulations is to teach the learning

We are attracted to the spirit of God he is though out the creation were are elders are found watching[the watchers] energy and matter cannot be created or destroyed they are sealed as one forever the 144,000 the seal has been broken we are real force creation and destroy are illusions to us what appear to be we deal with mathematically the war is a divine of mind God is not a man of war the fight is in the head the armor of the mind

Truth

Right

Peace

Foremost of faith

Freedom

Spirit

The heaven was harder to create then man by the fact it was made for us created after God being liken God a creation for us the sun moon and stars were created for us we are from God the creation is made for us made better then us for us you must know the creating at its best is for you the devil was easy to make he comes from us we come from the creating they have a father devil they eat with we eat with God made all right things for us and what God made for us we own the creation is made to give us what is ours we are made to be given reward by are walk and who we walk with righteous men are Gods evil men are devils in are physical body is a sun a light that shine the body full by the giving of the universe the more we receive the more you love what the universe is giving you the more you will be given a love made peace with the universe the universe will be at peace with you creation is a illusion of what is to come being the real of what will be made to come after the illusion is dead we will know longer be blind to the illusion of create and destroy we must see beyond the illusion of a heaven and the clock work of which the worlds move waste no time looking to the stars know you are sun moon and star know them as self yet waste no time the 10% follow the stars the knowledge wisdom and understanding of the 85% mass is what makes them dumb they think they know yet there teachings are from the 10% this is why they are a mental dead and cant move on past the freedom in emotion they are limited to there own self taught other then self

They are in a poor condition the 5% came to grow and develop as many of the 85% as we can to the 5% the devil cannot be righteous it is not in his make the righteous cannot be evil it is not created for him to be evil there is no thing as evil as the father devil of all devil the worst at evil it is not in his create to be of a devilishment as the devil was made we are pro-create the devil is of a make made to do what he do the voice of God is in you the word is in you God gave the world to us some knowledge not God cant find God in there truth those who know not God will be found poor in spirit the world in us is larger then the world on the out side of all created is liken to the supreme being divine goods divine evil creation comes form us for us God holds us up the devil keeps you on your knees in fear of a God we have never seen taught by the 10% we will never see by the so-called God they profess to be invisible that cannot be seen and will never be seen this is the teaching of the 10% star gazers to the 85% mass sleep to knowledge other then there own self were God is found yet they profess to wise the devil plays the part of angel the devil plays the part of the lion of Judah the Gods of planet earth are the original man a web to the spider is a universe the spider is the believer the web attract all in contact yet the spider cannot be trapped by its own web [universe] as the spider is the[believer] temptations are temperatures

Hot
Lukewarm
Cold
When the creation is

Nice
Sunny
Warm
because the sun is bright at the time there is a rope the woman holds the umbilical cord on the other side of the cord is devils in a world of sin women must pull the rope at here best

be calm
be peaceful
be at rest
be silent
no intoxicants
be of no evil

no man can love a woman as a child loves mother the child is not protected from devil with out mother who must pull harder then the devil if woman is reckless the baby will be shaped in sin born in sin yet a woman at peace with the womb the child is born in to sin born to a sinful world yet the child is not born with sin we are born in to a world of sin woman must hold the rope the way of life is the way of God the way of God is science and mathematics of the universe I shall be what is what I shall be I shall be and before it comes to past it will be spoken this wisdom cannot be taught with out the knowledge foundation of self the wise words are very near in your brain to your mouth we have built the west yet are story has been written out of the books history is not the black mans story are story is the real story

the spirit of God is

a personality
a characteristic
a temperament
a feature
a trait
a attitude
a custom
a mind statement
a mind set

we exist by the fact we have exodus to get here we exodus when we leave we come with a genesis we return with the genesis the geno-code of jah the letter of the law the angel Michael was a create by God for the physical and spiritual war the angel Gabriel was a create by God sent to give spiritual reward to mothers and the prophsts he came to Daniel with message he came to the mother mary and made a divine son who gave massage he came to the prophet muhammad on a high place of a mountain

the stars being infinite eternal absolute come from the soup of cosmic slop a disorder which we as alllahs nation gave alinement we are older then the stars so we waste no time looking to the sky as the 10% being the teachers of the mass 85% the mental of dumb who feel they know yet they know not the 5% are civilized and teach the knowledge of self in the man woman and child not the teaching of the 10% a God we cannot see until we die and pass on yet we as the 5% get now what we need before we die and that is knowledge of self not the stars the God we are is not invisible he is seen and heard every where we waste no time other then self this self is the time we utilize master the illusion of time and master build in the minute second of God theology of time we are on time as a twinkling of the eye which was[the bomb] the big bang which is a very long time before stars a long time before the slop of disorder it was not something we would want to see yet there is a change written in the cosmic

perfect
better
best

are soul is a geno the genesis before there was a creation knowledge of self was the creater knowledge self and create what you see it is your choice of which is your circle you have made reward or punishment is the justice you own we were made afraid we were made to doubt are place in the world as the greatest people to walk the space and time of the universe we were made to disbelieve are own heaven is at the feet of the mother the child is on all fours at his mothers feet being the heaven of the child we were knit with in are mothers womb we were woven together with in the bodys of woman we become her she becomes us we are one with woman kind[wisdom] she is one with us the child is most close to God in the blood of the mothers belly God is most close to the black man then the blood of the body it takes a mass amount of force to impregnate man can guess yet the woman knows her time mommies baby daddys maybe by the Jesus the five books of moses are born again not a flesh and blood law not a new law a add on to the old that cannot be broken father is grater then I which is 1 we are more then 1 we are us and we God is us and we grater then I knowledge is the foundation of all the Jesus taught the make of a wise flock the Jesus in his mind equality with God was not robbery this is the mind we must have the Jesus is a small God from a big God a small God to a create a big people from the original man came the Asiatic black man we are a create to act out

rulers
nobles
masters
legends
kings
lords
deity
God

the dome of are brain is were heaven is found

the heaven of wisdom
the heaven of freedom
the heaven of kingdom

we were dead as a people yet know we are alive in line to the God we are God knew before there was anything to know before there was a here and
before are mothers womb we were traveling space and time with God what you observe and reflect on is what you see not what appears to the blind eye what the naked eye sees musts be refined and purified with time prayer is good yet to mirror God is best those who see ahead of them self are those who se e create your mind and this create you have created will become a universe a man must be wise to his woman a woman must be wise with all men the in the old books which will remain old there is none old
there is prophesy of kings writing books in the future which is here and now fulfilled prophecy

men of philosophy
men of religion
men of science
men of music recordings

poverty is the state of the earth mass they are a poor people they are not informed with what we must know as civilized people the 10% teach the grave the 5% knows there is mental life beyond grave the 10% rulers live off the resource of the globe which is the original mans home the globe is are home the 10% use monetary laws to keep the dollar king in order to control the mass earth being are people the 5% are the salt of earth today we have are flavor and cannot be played a fool we are teachers of the poor Islam is a awareness a culture of firm rich black soil this is the cultivation of the mind the working of the soil is a peace beyond understanding the child star the best part of the mind and the peace beyond child in the mind and then the womb of Islam the peace found is in the growth and development that child is peace as he or she grows mother and father are at peace the seed of the mind is thought the mind is a make fruit the seed of the womb is a make of fruit a woman must know when to open her self to a man we were created for the universe and the universe is created for us what we are given from God by the universe we must utilize and make the best of what God has given by the universe God cannot change we are the same as he was

past

present

future

we are in the now we are in the old there is nothing new God is who he has been here before the infinite stars the Asiatic calendar starts at the trillions of years we live not in the world the world is in us a world a created by us for us

bless the eye with what is good
bless the ear with beautiful sound
bless the nose with sweet scents
bless the mouth with healthy foods

do to knowledge a woman is made big in belly a child is born moons circle planets the planets circle the sun the sun is a star that circle the galaxies all galaxies are in the universe is kept in place and order by the cosmos the 7th heaven makes us complete in the mind it is the completion of the 7 energy centers of the body what we see is a illusion with out observing and reflecting a righteous man must utilize time to see clear this is theology of time study what you see and make it clearly seen thought before you speak the physical has a chance to change we were a mental dead yet we came back the only constant is change self preservation is the first law energy cannot die the mind gives the spirit life if you know not God you are dead to God and others good for none the spirit in us is the light of genesis the four letter code of DNA the Y H W H letter of the old law jah is the spirit add on to the letter of the law the wise words of the Jesus we have the DNA of the 25 scientist who gave measure to the earth we share
DNA with them they are those who held sacred

The lion
The bull
The king
The eagle

The four letter geno code is Y H W H
The letter of the law jah the add on by the wise words of the Jesus by which woman

Call out with a cry the 24 scientists we hold there DNA they gave measure to the earth it is are turn to weigh the earth in its rights and wrongs the spirit makes the flesh fast and strong by the power of the spirit the flesh is made whole and healthy God made all things and in a meditative state he was done with his command to be God gave all the a created meditation to man who was given light of enlightenment from the light all men of light were separated from men of darkness the men of darkness were sent down under the low of the heavens the sun moon and star men of darkness were flying creatures made separate from the light creatures who all so fly and these beings would today be angels yet we know they were beings still healed sacred as God and Goddess the man of light is enlightened the man of darkness cannot overpower the light those of little faith must find balance in the earth those of little faith must move mountains which is a mass task the foremost of faith have no stumbling blocks we are a people who have measured the earth we move planets and stars by

Facts

Proofs

Times

Truths

Reason

Were there is light there is life were there life there is hue thought waves are hue colors of energy that effect thought

The yellow sun is a energy

The silver moon is a energy

Black space is a energy

The blue sky is a energy

The wooly sheep is a energy

Green vegetation is a energy

Gold is a energy

Brown is a energy

Fire red is a energy

Thought is life thought is light thought is energy thought is color what we see has effect on us yet we must sole control and be not affiliated be free of thought

The mind can become earthly

The mind can become soler

The mind can become universal

The mind can become resurrector

Some are not free to see God they cannot see there own self in the mirror in a good light we are formless God gives us form we are void and God fills the void by the Jesus love God love the heart love the mind love the soul love the neighborhood love self knowledge when we bomb on what is not ours the 4 devils who must be ran off are brain and murdered in the mind

We must bomb on lust
We must bomb on jealousy
We must bomb on envy
We must bomb on hate

These 4 devil heads must be destroyed in self so we can make are way in this pilgrimage of life in equal footing with the mass 85% some will wake to the 10% and some will wake from the poor condition of death made alive by the 5% yet it will be a small percent we are a mass of 4 billion 400 million 17 million original men 2 million west Indians making it 19 million chosen to civilization of the mass some will wake to the rules and regulations of the 5% the learning in knowledge is to be come God in a observe and a reflect in the mind and meet God not know God in name only

There are meany names and most names a of God are names held sacred names of the nature of the deity of God in the ranks God is first then others in there rank under the first matter needs energy to exist we are the spirit matter needs are spirit to live energy cannot die it cant be created it cannot be destroy matter will by the spirit will not be made to die some are in matter some are in the spirit some are in matter some are in spirit some are in matter and spirit those in the grave are of a mental dead in matter and knows not the spirit of them self are 85% those of the spirit have the power to civilize yet they lie and trick the 10% spirit is evil the matter and energy of the 5% is the mental and physical foods from which come are matter and energy the body needs matter and energy matters are the foods we eat energy is what we need to live we must eat mental we must eat physically what you eat is who you are God gave clean and unclean foods

To eat clean foods is the God degree you are what you eat to eat unclean foods is to act out ignorance and play dumb your thoughts are levels of light not colors we are colorful in mind by energy the levels of energy which are thoughts with effect on the mental colors are not just

A hue
A pigment
A tint
A tone
A shadow

They affect the mental they are a forms of energy it is not what is up in the sky it is what is down to earth [woman] we must care for the earth she is are home the womb of Islam we will die for are brothers and sisters before are word shall fail we are the God children of God yet we are not men of God like we should be we will fall for are people to wake them as a prince in all his glory he will die before his kingdom shall fall satan and Michael war over us in the high places is a war for the minds and bodys of are people Gabriel made us a- alike God by the son in ranks the Jesus Michael and satan in there war the spoils of this war is are women and are children in this last day a Christ has come to the lost found in the person of God and his messenger God is a magnetic pull a force of energy a light of thought waves as the5% we must have bread for the hungry self comes before others this is the first law yet we will die for are sisters and brothers before are word shell fail the circumference of the earth is the measurement of the earths goods and evils God is above the woman is above the child we must not only create self to know self we must re-create day to day which is the add and subtract mathematically day to day by the lessons one will see his life for the day being the science and mathematics of the universe by day to day we re-create the add on to the old there is none new all new is a add on to the old the tree of life is clean foods the tree of knowledge is unclean foods the

Pig is swine the serpent of worms with or with out the mass85% we cannot be played a fool the wisdom of are words will live on forever in the genesis there is a re-creantion of the stars

There is a re- creantion of the wildlife creatures

There is a re-creantion of believers made Gods do to knowledge we are God in the knowledge of self the only way to heaven is the feet of the woman we must bow to her out of respect for her mind and not just to her body satan would not bow to adam out of which was a respect bow not to your woman is to act as satan to adam we must pray to God for the womans perfections what makes a woman better then the next woman and a woman that is most best for you man and woman must be well versed in us and we God does not prey to you he has no need to prey to you we are free in mind God is not a dictator he does not force us he is a guide if we want guidance ask and you will receive Abraham came to the canaanites to make the wrongs of ham right he was humbled in Egypt given a sur name ham a queen of where came a tribe of kings studied ressrrection of the God Osiris studied the sun moon and stars as to which one was best being a Hebrew his children were given home in the black lands of Kemet Israelites were children of Abraham they were made to step down to yacub[Jacob] and the Israelites became children of yacub the children of Israel Hebrews a few devils are called yet only a few are chosen for the 5% this is a spirit of the mind the highest fight is in the mind the mind is in a constant struggle with the highs we have a speck in are eye we see the small things the devil has a plank in his eye his eye is on all things a greed for everything